D1619336

Reviews of Nonlinear Dynamics and Complexity

Edited by
Heinz Georg Schuster

Related Titles

N. Santoro

Design and Analysis of Distributed Algorithms

2006

ISBN 978-0-471-71997-8

H.P. Alesso, C.F. Smith

Thinking on the Web
Berners-Lee, Gödel and Turing

2008

ISBN 978-0-470-11881-4

D. Ford, L.-E. Gadde, H. Håkansson, I. Snehota

The Business Marketing Course
Managing in Complex Networks

2006

ISBN 978-0-470-03450-7

A. von Meier

Electric Power Systems
A Conceptual Introduction

2006

ISBN 978-0-471-17859-0

H. Eisner

Managing Complex Systems
Thinking Outside the Box

2005

ISBN 978-0-471-69006-1

F.C. Moon

Chaotic Vibrations
An Introduction for Applied Scientists and Engineers

2004

ISBN 978-0-471-67908-0

L.V. Yakushevich

Nonlinear Physics of DNA

2004

ISBN 978-3-527-40417-9

M. Kantardzic

Data Mining
Concepts, Models, Methods, and Algorithms

2005

ISBN 978-0-471-65605-0

Reviews of Nonlinear Dynamics and Complexity

Edited by
Heinz Georg Schuster

WILEY-VCH Verlag GmbH & Co. KGaA

The Editor

Heinz Georg Schuster
University of Kiel
schuster@theo-physik.uni-kiel.de

Library of Congress Card No.: applied for

British Library Cataloguing-in-Publication Data
A catalogue record for this book is available from the British Library.

Bibliographic information published by the Deutsche Nationalbibliothek
Die Deutsche Nationalbibliothek lists this publication in the Deutsche Nationalbibliografie; detailed bibliographic data are available on the Internet at http://dnb.d-nb.de

Printed in the Federal Republic of Germany
Printed on acid-free paper

Typesetting Uwe Krieg, Berlin
Printing betz-druck GmbH, Darmstadt
Bookbinding Litges & Dopf GmbH, Heppenheim

ISBN: 978-3-527-40729-3

Contents

Review of Nonlinear Dynamics and Complexity. Edited by Heinz Georg Schuster

ISBN: 978-3-527-40729-3

Preface

Nonlinear behavior is ubiquitous in nature and ranges from fluid dynamics, via neural and cell dynamics to the dynamics of financial markets. The most prominent feature of nonlinear systems is that small external disturbances can induce large changes in their behavior. This can and has been used for effective feedback control in many systems, from Lasers to chemical reactions and the control of nerve cells and heartbeats. A new hot topic are nonlinear effects that appear on the nanoscale. Nonlinear control of the atomic force microscope has improved it accuracy by orders of magnitude. Nonlinear electromechanical oscillations of nano-tubes, turbulence and mixing of fluids in nano-arrays and nonlinear effects in quantum dots are further examples.

Complex systems consist of large networks of coupled nonlinear devices. The observation that scalefree networks describe the behavior of the internet, cell metabolisms, financial markets and economic and ecological systems has lead to new findings concerning their behavior, such as damage control, optimal spread of information or the detection of new functional modules, that are pivotal for their description and control.

This shows that the field of *Nonlinear Dynamics and Complexity* consists of a large body of theoretical and experimental work with many applications, which is nevertheless governed and held together by some very basic principles, such as control, networks and optimization. The individual topics are definitely interdisciplinary which makes it difficult for researchers to see what new solutions – which could be most relevant for them – have been found by their scientific neighbors. Therefore its seems quite urgent to provide *Reviews of Nonlinear Dynamics and Complexity* where researchers or newcomers to the field can find the most important recent results, described in a fashion which breaks the barriers between the disciplines.

In this first volume topics range from nano-mechanical oscillators, via random Boolean networks and control to extreme climatic and seismic events. I would like to thank all authors for their excellent contributions. If readers take from these interdisciplinary reviews some inspiration for their further research this volume would fully serve its purpose.

Review of Nonlinear Dynamics and Complexity. Edited by Heinz Georg Schuster

ISBN: 978-3-527-40729-3

I am grateful to all members of the Editorial Board and the staff from Wiley-VCH for their excellent help and would like to invite my colleagues to contribute to the next volumes.

Kiel, February 2008 *Heinz Georg Schuster*

List of Contributors

Armin Bunde
Institut für Theoretische Physik III
Justus-Liebig-Universität Giessen
Heinrich-Buff-Ring 16
35392 Giessen
Germany
armin.bunde@physik.uni-giessen.de

M. C. Cross
Department of Physics
California Institute of Technology
Pasadena, CA 91125
USA
mcc@caltech.edu

Barbara Drossel
Institut für Festkörperphysik
Technische Universität Darmstadt
Hochschulstrasse 6
64289 Darmstadt
Germany
barbara.drossel@physik.tu-darmstadt.de

Jan F. Eichner
Institut für Theoretische Physik III
Justus-Liebig-Universität Giessen
Heinrich-Buff-Ring 16
35392 Giessen
Germany
jan.f.eichner@physik.uni-giessen.de

Bernold Fiedler
Institut für Mathematik I
FU Berlin
Arnimallee 2–6
14195 Berlin
Germany
fiedler@math.fu-berlin.de

Valentin Flunkert
Institut für Theoretische Physik
Technische Universität Berlin
Hardenbergstrasse 36
10623 Berlin
Germany
tino@physik.tu-berlin.de

Marc Georgi
Institut für Mathematik I
FU Berlin
Arnimallee 2–6
14195 Berlin
Germany
georgi@inf.fu-berlin.de

Bailin Hao
The Institute of Theoretical Physics
55 Zhongguancun East Road
Haidan District
Beijing 100080
China
and
The T-Life Research Center
Fudan University
220 Handan Road
Shanghai 200433
China
hao@mail.itp.ac.cn

Shlomo Havlin
Minerva Center and Department
of Physics
Bar-Ilan University
Ramat-Gan
Israel
havlin@ophir.ph.biu.ac.il

Review of Nonlinear Dynamics and Complexity. Edited by Heinz Georg Schuster

ISBN: 978-3-527-40729-3

Philipp Hövel
Institut für Theoretische Physik
Technische Universität Berlin
Hardenbergstrasse 36
10623 Berlin
Germany
phoevel@physik.tu-berlin.de

Jan W. Kantelhardt
Institut für Physik
Martin-Luther-Universität
Halle-Wittenberg
Germany
jan.kantelhardt@physik.uni-halle.de

Sabine Lennartz
Institut für Theoretische Physik III
Justus-Liebig-Universität Giessen
Heinrich-Buff-Ring 16
35392 Giessen
Germany
sabine.lennartz@physik.uni-giessen.de

Ron Lifshitz
Raymond and Beverly Sackler School of
Physics and Astronomy
Tel Aviv University
Tel Aviv 69978
Israel
lifshitz@caltech.edu

Arkady Pikovsky
Department of Physics
Potsdam University
Am Neuen Palais 10
14469 Potsdam
Germany
pikovsky@stat.physik.uni-potsdam.de

Michael Rosenblum
Department of Physics
Potsdam University
Am Neuen Palais 10
14469 Potsdam
Germany
mRos@agnld.uni-potsdam.de

Eckehard Schöll
Institut für Theoretische Physik
Technische Universität Berlin
Hardenbergstrasse 36
10623 Berlin
Germany
schoell@physik.tu-berlin.de

Huimin Xie
Department of Mathematics
Suzhou University
Suzhou 215006
China
szhmxie@yahoo.com.cn

1
Nonlinear Dynamics of Nanomechanical and Micromechanical Resonators

Ron Lifshitz and M. C. Cross

1.1
Nonlinearities in NEMS and MEMS Resonators

In the last decade we have witnessed exciting technological advances in the fabrication and control of microelectromechanical and nanoelectromechanical systems (MEMS & NEMS) [1–5]. Such systems are being developed for a host of nanotechnological applications, such as highly-sensitive mass [6–8], spin [9], and charge detectors [10, 11], as well as for basic research in the mesoscopic physics of phonons [12], and the general study of the behavior of mechanical degrees of freedom at the interface between the quantum and the classical worlds [13, 14]. Surprisingly, MEMS & NEMS have also opened up a whole new experimental window into the study of the nonlinear dynamics of discrete systems in the form of nonlinear micromechanical and nanomechanical oscillators and resonators.

The purpose of this review is to provide an introduction to the nonlinear dynamics of micromechanical and nanomechanical resonators that starts from the basics, but also touches upon some of the advanced topics that are relevant for current experiments with MEMS & NEMS devices. We begin in this section with a general motivation, explaining why nonlinearities are so often observed in NEMS & MEMS devices. In § 1.2 we describe the dynamics of one of the simplest nonlinear devices – the Duffing resonator – while giving a tutorial in secular perturbation theory as we calculate its response to an external drive. We continue to use the same analytical tools in § 1.3 to discuss the dynamics of a parametrically-excited Duffing resonator, building up to the description of the dynamics of an array of coupled parametrically-excited Duffing resonators in § 1.4. We conclude in § 1.5 by giving an amplitude equation description for the array of coupled Duffing resonators, allowing us to extend our analytic capabilities in predicting and explaining the nature of its dynamics.

Review of Nonlinear Dynamics and Complexity. Edited by Heinz Georg Schuster

ISBN: 978-3-527-40729-3

1.1.1 Why Study Nonlinear NEMS and MEMS?

Interest in the nonlinear dynamics of microelectromechanical and nanoelectromechanical systems (MEMS & NEMS) has grown rapidly over the last few years, driven by a combination of practical needs as well as fundamental questions. Nonlinear behavior is readily observed in micro- and nano-scale mechanical devices [3, 15–32]. Consequently, there exists a practical need to understand this behavior in order to avoid it when it is unwanted, and exploit it efficiently when it is wanted. At the same time, advances in the fabrication, transduction, and detection of MEMS & NEMS resonators has opened up an exciting new experimental window into the study of fundamental questions in nonlinear dynamics. Typical nonlinear MEMS & NEMS resonators are characterized by extremely high frequencies – recently going beyond 1 GHz [33–35] – and relatively weak dissipation, with quality factors in the range of 10^2–10^4. For such devices the regime of physical interest is that of steady state motion as transients tend to disappear before they are detected. This, and the fact that weak dissipation can be treated as a small perturbation, provide a great advantage for quantitative theoretical study. Moreover, the ability to fabricate arrays of tens to hundreds of coupled resonators opens new possibilities in the study of nonlinear dynamics of intermediate numbers of degrees of freedom – much larger than one can study in macroscopic or table-top experiments, yet much smaller than one studies when considering nonlinear aspects of phonon dynamics in a crystal. The collective response of coupled arrays might be useful for signal enhancement and noise reduction [36], as well as for sophisticated mechanical signal processing applications. Such arrays have already exhibited interesting nonlinear dynamics ranging from the formation of extended patterns [37] – as one commonly observes in analogous continuous systems such as Faraday waves – to that of intrinsically localized modes [38–40]. Thus, nanomechanical resonator arrays are perfect for testing dynamical theories of discrete nonlinear systems with many degrees of freedom. At the same time, the theoretical understanding of such systems may prove useful for future nanotechnological applications.

1.1.2 Origin of Nonlinearity in NEMS and MEMS Resonators

We are used to thinking about mechanical resonators as being simple harmonic oscillators, acted upon by linear elastic forces that obey Hooke's law. This is usually a very good approximation as most materials can sustain relatively large deformations before their intrinsic stress-strain relation breaks away from a simple linear description. Nevertheless, one commonly encounters nonlinear dynamics in micromechanical and nanomechanical resonators long before the intrinsic nonlinear regime is reached. Most evident are nonlinear effects that enter the equation of motion in the form of a force that is proportional to the cube of the displacement αx^3. These turn a simple harmonic resonator with a linear restoring force into a so-called Duffing resonator. The two main origins of the observed nonlinear effects are illustrated below with the help of two typical examples. These are due to the effect of external potentials that are often nonlinear, and geometric effects that introduce nonlinearities even though the individual forces that are involved are all linear. The Duffing nonlinearity αx^3 can be positive, assisting the linear restoring force, making the resonator stiffer, and increasing its resonance frequency. It can also be negative, working against the linear restoring force, making the resonator softer, and decreasing its resonance frequency. The two examples we give below illustrate how both of these situations can arise in realistic MEMS & NEMS devices.

Additional sources of nonlinearity may be found in experimental realizations of MEMS and NEMS resonators due to practical reasons. These may include nonlinearities in the actuation and in the detection mechanisms that are used for interacting with the resonators. There could also be nonlinearities that result from the manner in which the resonator is clamped by its boundaries to the surrounding material. These all introduce external factors that may contribute to the overall nonlinear behavior of the resonator.

Finally, nonlinearities often appear in the damping mechanisms that accompany every physical resonator. We shall avoid going into the detailed description of the variety of physical processes that govern the damping of a resonator. Suffice it to say that whenever it is reasonable to expand the forces acting on a resonator up to the cube of the displacement x^3, it should correspondingly be reasonable to add to the linear damping which is proportional to the velocity of the resonator $\dot{x}$, a nonlinear damping term of the form $x^2\dot{x}$, which increases with the amplitude of motion. Such nonlinear damping will be considered in our analysis below.

1.1.3
Nonlinearities Arising from External Potentials

As an example of the effect of an external potential, let us consider a typical situation, discussed for example by Cleland and Roukes [10,11], and depicted in Fig. 1.1, in which a harmonic oscillator is acted upon by an external electrostatic force. This could be implemented by placing a rigid electrically-charged base electrode near an oppositely-charged NEMS or MEMS resonator. If the equilibrium separation between the resonator and the base electrode in the absence of electric charge is d, the deviation away from this equilibrium position is denoted by X, the effective elastic spring constant of the resonator is K,

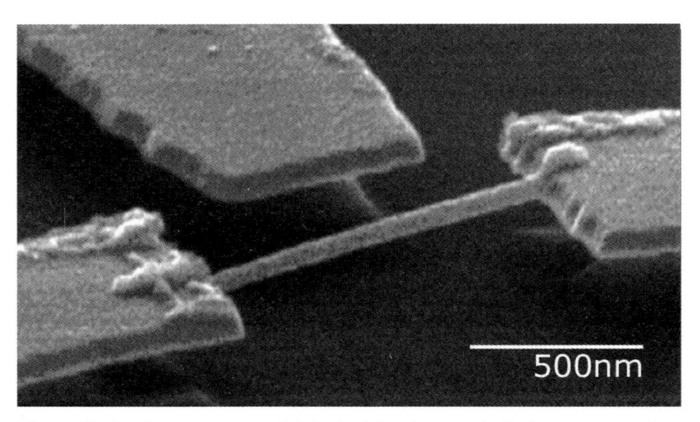

Fig. 1.1 A 43 nanometer thick doubly-clamped platinum nanowire with an external electrode that can be used to tune its natural frequency as well as its nonlinear properties. (Adapted with permission from [30]).

and the charge q on the resonator is assumed to be constant, then the potential energy of the resonator is given by

$$V(X) = \frac{1}{2}KX^2 - \frac{C}{d+X}\,. \tag{1.1}$$

In SI units $C = Aq^2/4\pi\epsilon_0$, where A is a numerical factor of order unity that takes into account the finite dimensions of the charged resonator and base electrode. The new equilibrium position X_0 in the presence of charge can be determined by solving the cubic equation

$$\frac{dV}{dX} = KX + \frac{C}{(d+X)^2} = 0. \tag{1.2}$$

If we now expand the potential acting on the resonator in a power series in the deviation $x = X - X_0$ from this new equilibrium we obtain

$$\begin{aligned} V(x) &\simeq V(X_0) + \frac{1}{2}\left(K - \frac{2C}{(d+X_0)^3}\right)x^2 + \frac{C}{(d+X_0)^4}x^3 - \frac{C}{(d+X_0)^5}x^4 \\ &= V(X_0) + \frac{1}{2}kx^2 + \frac{1}{3}\beta x^3 + \frac{1}{4}\alpha x^4. \end{aligned} \tag{1.3}$$

This gives rise, without any additional driving or damping, to an equation of motion of the form

$$m\ddot{x} + kx + \beta x^2 + \alpha x^3 = 0, \qquad \text{with} \quad \beta > 0,\ \alpha < 0, \tag{1.4}$$

where m is the effective mass of the resonator; k is the new effective spring constant, which is softened by the electrostatic attraction to the base electrode, but note that if $2C/(d+X_0)^3 > K$, the electrostatic force exceeds the elastic restoring force and the resonator is pulled onto the base electrode; β is a positive symmetry-breaking quadratic elastic constant that pulls the resonator towards the base electrode regardless of the sign of x; and α is the cubic, or Duffing, elastic constant that owing to its negative sign softens the effect of the linear restoring force. It should be sufficient to stop the expansion here, unless the amplitude of the motion is much larger than the size of the resonator, or if by some coincidence the effects of the quadratic and cubic nonlinearities happen to cancel each other out – a situation that will become clearer after reading § 1.2.3.

1.1.4 Nonlinearities Due to Geometry

As an illustration of how nonlinearities can emerge from linear forces due to geometric effects, consider a doubly-clamped thin elastic beam, which is one of the most commonly encountered NEMS resonators. Because of the clamps at both ends, as the beam deflects in its transverse motion it necessarily stretches. As long as the amplitude of the transverse motion is much smaller than the width of the beam this effect can be neglected. But with NEMS beams it is often the case that they are extremely thin, and are driven quite strongly, making it common for the amplitude of vibration to exceed the width. Let us consider this effect in some detail by starting with the Euler–Bernoulli equation, which is the commonly used approximate equation of motion for a thin beam [41]. For a transverse displacement $X(z,t)$ from equilibrium, which is much smaller than the length L of the beam, the equation is

$$\rho S\frac{\partial^2 X}{\partial t^2} = -EI\frac{\partial^4 X}{\partial z^4} + T\frac{\partial^2 X}{\partial z^2}, \tag{1.5}$$

where z is the coordinate along the length of the beam; ρ the mass density; S the area of the cross section of the beam; E the Young modulus; I the moment of inertia; and T the tension in the beam, which is composed of its inherent tension T_0 and the additional tension ΔT due to its bending that induces an extension ΔL in the length of the beam. Inherent tension results from the fact that in equilibrium in the doubly-clamped configuration, the actual length of the beam may differ form its rest length being either extended (positive T_0) or compressed (negative T_0). The additional tension ΔT is given by the strain, or relative extension of the beam $\Delta L/L$ multiplied by Young's modulus E and the area of the beam's cross section S. For small displacements the total length of the beam can be expanded as

$$L + \Delta L = \int_0^L dz \sqrt{1 + \left(\frac{\partial X}{\partial z}\right)^2} \simeq L + \frac{1}{2}\int_0^L dz \left(\frac{\partial X}{\partial z}\right)^2. \tag{1.6}$$

The equation of motion (1.5) then clearly becomes nonlinear

$$\rho S \frac{\partial^2 X}{\partial t^2} = -EI\frac{\partial^4 X}{\partial z^4} + \left[T_0 + \frac{ES}{2L}\int_0^L dz \left(\frac{\partial X}{\partial z}\right)^2\right]\frac{\partial^2 X}{\partial z^2}. \tag{1.7}$$

We can treat this equation perturbatively [42,43]. We consider first the linear part of the equation, which has the form of Eq. (1.5) with T_0 in place of T, separate the variables,

$$X_n(z,t) = x_n(t)\phi_n(z), \tag{1.8}$$

and find its spatial eigenmodes $\phi_n(z)$, where we use the convention that the local maximum of the eigenmode $\phi_n(z)$ that is nearest to the center of the beam is scaled to 1. Thus $x_n(t)$ measures the actual deflection at the point nearest to its center that extends the furthest. Next, we assume that the beam is vibrating predominantly in one of these eigenmodes and use this assumption to evaluate the effective Duffing parameter α_n, multiplying the x_n^3 term in the equation of motion for this mode. Corrections to this approximation will appear only at higher orders of x_n. We multiply Eq. (1.7) by the chosen eigenmode $\phi_n(z)$, and integrate over z to get, after some integration by parts, a Duffing equation of motion for the amplitude of the nth mode $x_n(t)$,

$$\ddot{x}_n + \left[\frac{EI}{\rho S}\frac{\int \phi_n''^2 dz}{\int \phi_n^2 dz} + \frac{T_0}{\rho S}\frac{\int \phi_n'^2 dz}{\int \phi_n^2 dz}\right]x_n + \left[\frac{E}{2\rho L}\frac{\left(\int \phi_n'^2 dz\right)^2}{\int \phi_n^2 dz}\right]x_n^3 = 0, \tag{1.9}$$

where primes denote derivatives with respect to z, and all the integrals are from 0 to L. Note that we have obtained a positive Duffing term, indicating a stiffening nonlinearity, as opposed to the softening nonlinearity that we saw in the previous section. Also note that the effective spring constant can be made negative by compressing the equilibrium beam, thus making T_0 large and negative. This may lead to the so-called Euler instability, which is a buckling instability of the beam.

To evaluate the effective Duffing nonlinearity α_n for the nth mode, we introduce a dimensionless parameter $\hat{\alpha}_n$ by rearranging the equation of motion (1.9), to have the form

$$\ddot{x}_n + \omega_n^2 x_n\left[1 + \hat{\alpha}_n \frac{x_n^2}{d^2}\right] = 0, \tag{1.10}$$

where ω_n is the normal frequency of the nth mode, d is the width, or diameter, of the beam in the direction of the vibration, and x_n is the maximum displacement of the beam near its center. This parameter can then be evaluated regardless of the actual dimension of the beam.

In the limit of small residual tension T_0 the eigenmodes are those dominated by bending given by [41]

$$\phi_n(z) = \frac{1}{a_n}\Big[(\sin k_n L - \sinh k_n L)(\cos k_n z - \cosh k_n z) - (\cos k_n L - \cosh k_n L)(\sin k_n z - \sinh k_n z)\Big], \tag{1.11}$$

where a_n is the value of the function in the square brackets at its local maximum that is closest to $z = 0.5$, and the wave vectors k_n are solutions of the transcendental equation $\cos k_n L \cosh k_n L = 1$. The first few values are

$$\{k_n L\} \simeq \{4.7300, 7.8532, 10.9956, 14.1372, 17.2788, 20.4204\ldots\}, \tag{1.12}$$

and the remaining ones tend towards odd-integer multiples of $\pi/2$, as n increases. Using these eigenfunction we can obtain explicit values for the dimensionless Duffing parameters for the different modes, by calculating

$$\hat{\alpha}_n = \frac{Sd^2}{2I}\frac{\left(\frac{1}{L}\int \phi_n'^2 dz\right)^2}{\frac{1}{L}\int \phi_n''^2 dz} \equiv \frac{Sd^2}{2I}\hat{\beta}_n. \tag{1.13}$$

The first few values are

$$\{\hat{\beta}_n\} \simeq \{0.1199, 0.2448, 0.3385, 0.3706, 0.3908, 0.4068, 0.4187, \ldots\}, \tag{1.14}$$

tending to an asymptotic value of $1/2$ as $n \to \infty$. For beams with rectangular or circular cross sections, the geometric prefactor evaluates to

$$\frac{Sd^2}{2I} = \begin{cases} 16 & \text{Circular cross section} \\ 6 & \text{Rectangular cross section} \end{cases} \tag{1.15}$$

Thus the dimensionless Duffing parameters are of order 1, and therefore the significance of the nonlinear behavior is solely determined by the ratio of the deflection to the width of the beam.

In the limit of large equilibrium tension, the beam essentially behaves as a string with relatively negligible resistance to bending. The eigenmodes are those of a string,

$$\phi_n(z) = \sin\left(\frac{n\pi}{L}z\right), \quad n = 1, 2, 3, \ldots, \tag{1.16}$$

and, if we denote the equilibrium extension of the beam as $\Delta L_0 = LT_0/ES$, the dimensionless Duffing parameters are exactly given by

$$\hat{\alpha}_n = \frac{d^2}{2\Delta L_0} \int \phi_n'^2 dz = \frac{(n\pi d)^2}{4L\Delta L_0}. \tag{1.17}$$

In the large tension limit, as in the case of a string, the dimensionless Duffing parameters are proportional to the inverse aspect ratio of the beam d/L times the ratio between its width and the extension from its rest length $d/\Delta L_0$, at least one of which can be a very small parameter. For this reason nonlinear effects are relatively negligible in these systems.

1.2
The Directly-driven Damped Duffing Resonator

1.2.1
The Scaled Duffing Equation of Motion

Let us begin by considering a single nanomechanical Duffing resonator with linear and nonlinear damping that is driven by an external sinusoidal force. We shall start with the common situation where there is symmetry between x and $-x$, and consider later the changes that are introduced by adding symmetry-breaking terms. Such a resonator is described by the equation of motion

$$m\frac{d^2\tilde{x}}{d\tilde{t}^2} + \Gamma\frac{d\tilde{x}}{d\tilde{t}} + m\omega_0^2\tilde{x} + \tilde{\alpha}\tilde{x}^3 + \tilde{\eta}\tilde{x}^2\frac{d\tilde{x}}{d\tilde{t}} = \tilde{G}\cos\tilde{\omega}\tilde{t}, \tag{1.18}$$

where m is its effective mass, $k = m\omega_0^2$ is its effective spring constant, $\tilde{\alpha}$ is the cubic spring constant, or Duffing parameter, Γ is the linear damping rate, and $\tilde{\eta}$ is the coefficient of nonlinear damping – damping that increases with the amplitude of oscillation. We follow the convention that physical parameters that are immediately rescaled appear with twidles, as the first step in dealing with such an equation is to scale away as many unnecessary parameters as possible, leaving only those that are physically-significant, thus removing all of the twidles. We do so by: (1) Measuring time in units of ω_0^{-1} so that the dimensionless time variable is $t = \omega_0\tilde{t}$. (2) Measuring amplitudes of motion in units of length for which a unit-amplitude oscillation doubles the frequency of the resonator. This is achieved by taking the dimensionless length variable to be $x = \tilde{x}\sqrt{\tilde{\alpha}/m\omega_0^2}$. For the doubly-clamped beam of width or diameter d, discussed in § 1.1.4, this length is $x = \tilde{x}\sqrt{\hat{\alpha}_n}/d$. (3) Dividing the equation by an overall factor of $\omega_0^3\sqrt{m^3/\tilde{\alpha}}$. This yields a scaled Duffing equation of the form

$$\ddot{x} + Q^{-1}\dot{x} + x + x^3 + \eta x^2 \dot{x} = G\cos\omega t, \tag{1.19}$$

where dots denote derivatives with respect to the dimensionless time t, all the dimensionless parameters are related to the physical ones by

$$Q^{-1} = \frac{\Gamma}{m\omega_0}; \quad \eta = \frac{\tilde{\eta}\omega_0}{\tilde{\alpha}}; \quad G = \frac{\tilde{G}}{\omega_0^3}\sqrt{\frac{\tilde{\alpha}}{m^3}}; \quad \text{and} \quad \omega = \frac{\tilde{\omega}}{\omega_0}; \tag{1.20}$$

and Q is the quality factor of the resonator.

1.2.2
A Solution Using Secular Perturbation Theory

We proceed to calculate the response of the damped Duffing resonator to an external sinusoidal drive, as given by Eq. (1.19), by making use of secular perturbation theory [44,45]. We do so in the limit of a weak linear damping rate Q^{-1}, which we use to define a small expansion parameter, $Q^{-1} \equiv \epsilon \ll 1$. In most actual applications Q is at least on the order of 100, thus taking this limit is well-justified. We also consider the limit of weak oscillations where it is justified to truncate the expansion of the force acting on the resonator at the third power of x. We do so by requiring that the cubic force x^3 be a factor of ϵ smaller than the linear force, or equivalently, by requiring the deviation from equilibrium x to be on the order of $\sqrt{\epsilon}$. We ensure that the external driving force has the right strength to induce such weak oscillations by having it enter the equation at the same order as all the other physical effects. This turns out to require the amplitude of the drive to be $G = \epsilon^{3/2}g$. To see why, recall that for a regular linear resonance x is proportional to GQ. Here we want x to be of order $\sqrt{\epsilon}$, and Q is of order ϵ^{-1}, thus G has to be of order $\epsilon^{3/2}$. Finally, since damping is weak we expect to see response only close to the resonance frequency. We therefore take the driving frequency to be of the form $\omega = 1 + \epsilon\Omega$. The equation of motion (1.19) therefore becomes

$$\ddot{x} + \epsilon\dot{x} + x + x^3 + \eta x^2 \dot{x} = \epsilon^{3/2} g\cos(1 + \epsilon\Omega)t. \tag{1.21}$$

This is the equation we shall study using secular perturbation theory, while comparing from time to time back to the original physical equation (1.18).

Expecting the motion of the resonator away from equilibrium to be on the order of $\epsilon^{1/2}$, we try a solution of the form

$$x(t) = \frac{\sqrt{\epsilon}}{2}\left(A(T)e^{it} + c.c.\right) + \epsilon^{3/2}x_1(t) + \ldots, \tag{1.22}$$

where $c.c.$ denotes complex conjugation.

The lowest order contribution to this solution is based on the solution to the linear equation of motion of a simple harmonic oscillator (SHO) $\ddot{x} + x = 0$, where $T = \epsilon t$ is a slow time variable, allowing the complex amplitude $A(T)$ to vary slowly in time, due to the effect of all the other terms in the equation. As we shall immediately see, the slow temporal variation of $A(T)$ also allows us to ensure that the perturbative correction $x_1(t)$, as well as all higher-order corrections to the linear equation, do not diverge (as they do if one uses naive perturbation theory). Using the relation

$$\dot{A} = \frac{dA}{dt} = \epsilon \frac{dA}{dT} \equiv \epsilon A', \tag{1.23}$$

we calculate the time derivatives of the trial solution (1.22)

$$\dot{x} = \frac{\sqrt{\epsilon}}{2}\left([iA + \epsilon A']e^{it} + c.c.\right) + \epsilon^{3/2}\dot{x}_1(t) + \ldots \tag{1.24a}$$

$$\ddot{x} = \frac{\sqrt{\epsilon}}{2}\left([-A + 2i\epsilon A' + \epsilon^2 A'']e^{it} + c.c.\right) + \epsilon^{3/2}\ddot{x}_1(t) + \ldots \tag{1.24b}$$

By substituting these expressions back into the equation of motion (1.21), and picking out all terms of order $\epsilon^{3/2}$, we get the following equation for the first perturbative correction

$$\ddot{x}_1 + x_1 = \left(-iA' - i\frac{1}{2}A - \frac{3 + i\eta}{8}|A|^2 A + \frac{g}{2}e^{i\Omega T}\right)e^{it} - \frac{1 + i\eta}{8}A^3 e^{3it} + c.c. \tag{1.25}$$

The collection of terms proportional to e^{it} on the right-hand side of Eq. (1.25), called the secular terms, act like a force, driving the SHO on the left-hand side exactly at its resonance frequency. The sum of all these terms must therefore vanish so that the perturbative correction $x_1(t)$ will not diverge. This requirement is the so-called "solvability condition", giving us an equation for determining the slowly varying amplitude $A(T)$,

$$\frac{dA}{dT} = -\frac{1}{2}A + i\frac{3}{8}|A|^2 A - \frac{\eta}{8}|A|^2 A - i\frac{g}{2}e^{i\Omega T}. \tag{1.26}$$

This general equation could be used to study many different effects [46]. Here we use it to study the steady-state dynamics of the driven Duffing resonator.

We ignore initial transients, and assume that a steady-state solution of the form

$$A(T) = ae^{i\Omega T} \equiv |a|e^{i\phi}e^{i\Omega T} \tag{1.27}$$

exists. With this expression for the slowly varying amplitude $A(T)$ the solution to the original equation of motion (1.21) becomes an oscillation at the drive frequency $\omega = 1 + \epsilon\Omega$,

$$x(t) = \epsilon^{1/2}|a|\cos(\omega t + \phi) + O(\epsilon^{3/2}), \tag{1.28}$$

where we are not interested in the actual correction $x_1(t)$ of order $\epsilon^{3/2}$, but rather in finding the fixed complex amplitude a of the lowest order term. This amplitude a can be any solution of the equation

$$\left[\left(\frac{3}{4}|a|^2 - 2\Omega\right) + i\left(1 + \frac{\eta}{4}|a|^2\right)\right] a = g, \tag{1.29}$$

obtained by substituting the steady-state solution (1.27) into Eq. (1.26) of the secular terms.

The magnitude and phase of the response are then given explicitly by

$$|a|^2 = \frac{g^2}{\left(2\Omega - \frac{3}{4}|a|^2\right)^2 + \left(1 + \frac{1}{4}\eta|a|^2\right)^2}, \tag{1.30a}$$

and

$$\tan\phi = \frac{1 + \frac{1}{4}\eta|a|^2}{2\Omega - \frac{3}{4}|a|^2}. \tag{1.30b}$$

By reintroducing the original physical scales we can obtain the physical solution to the original equations of motion, $\tilde{x}(\tilde{t}) \simeq \tilde{x}_0\cos(\tilde{\omega}\tilde{t} + \phi)$, where $\tilde{x}_0 = |a|\sqrt{\Gamma\omega_0/\tilde{\alpha}}$, and therefore

$$\tilde{x}_0^2 = \frac{\left(\frac{\tilde{G}}{2m\omega_0^2}\right)^2}{\left(\frac{\tilde{\omega}-\omega_0}{\omega_0} - \frac{3}{8}\frac{\tilde{\alpha}}{m\omega_0^2}\tilde{x}_0^2\right)^2 + \left(\frac{1}{2}Q^{-1} + \frac{1}{8}\frac{\tilde{\eta}}{m\omega_0}\tilde{x}_0^2\right)^2} \tag{1.31a}$$

and

$$\tan\phi = \frac{\frac{\Gamma}{2} + \frac{\tilde{\eta}}{8}\tilde{x}_0^2}{m\tilde{\omega} - m\omega_0 - \frac{3\tilde{\alpha}}{8\omega_0}\tilde{x}_0^2}. \tag{1.31b}$$

The scaled response functions (1.30) are plotted in Fig. 1.2 for a drive with a scaled amplitude of $g = 3$ with and without nonlinear damping. The response without nonlinear damping is shown also in Fig. 1.3 for a sequence of increasing drive amplitudes ranging from $g = 0.1$, where the response is essentially

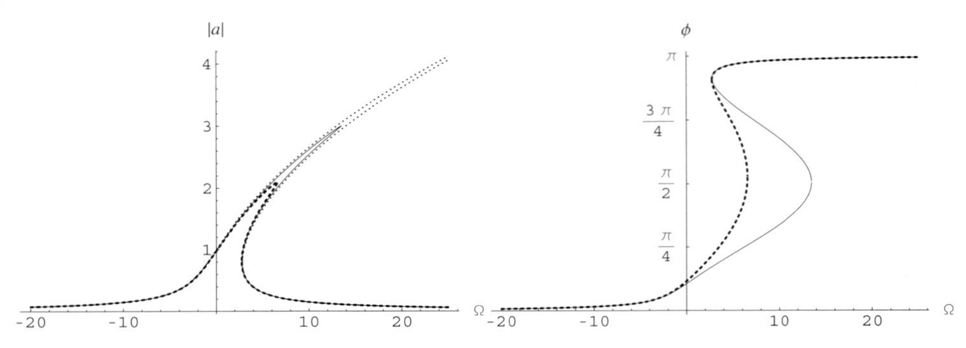

Fig. 1.2 Magnitude $|a|$ (left) and phase ϕ (right) of the response of a Duffing resonator as a function of the frequency Ω, for a fixed driving amplitude $g = 3$. The thin solid curves show the response without any nonlinear damping ($\eta = 0$). The thick dotted curves show the response with nonlinear damping ($\eta = 0.1$). The thin dotted curve on the left shows the response without any kind of damping [$Q^{-1} = 0$ and $\eta = 0$ in the original equation (1.19)]. The phase in this case is 0 along the whole upper-left branch and π along the whole lower-right branch, and so is not plotted on the right.

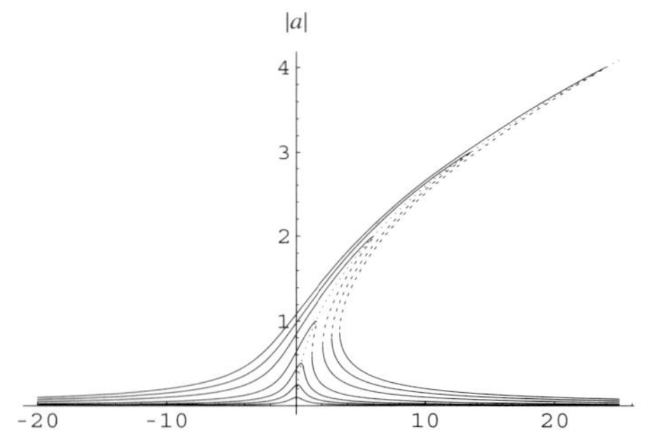

Fig. 1.3 Magnitudes $|a|$ (left) and phases ϕ (right) of the response of a Duffing resonator as a function of the frequency Ω, for a sequence of increasing values of the drive amplitude $0.1 \le g \le 4.0$, without nonlinear damping ($\eta = 0$). Solid curves indicate stable solutions of the response function (1.30), while dashed curves indicate unstable solutions.

linear, to the value of $g = 4$. Note that due to our choice of a positive Duffing nonlinearity the resonator becomes stiffer and its frequency higher as the amplitude increases. The response amplitude of the driven resonator therefore increases with increasing frequency, until it reaches a saddle-node bifurcation and drops abruptly to zero. A negative Duffing parameter would produce a mirror image of this response curve.

One sees that the magnitude of the response, given by Eq. (1.30a), formally-approaches the Lorentzian response of a linear SHO if we let the nonlinear terms in the original equation of motion tend to zero. Their existence modifies the response function with the appearance of the squared magnitude $|a|^2$ in the denominator on the right-hand side of Eq. (1.30a), turning the solution into a cubic polynomial in $|a|^2$. As such there are either one or three real solutions for $|a|^2$, and therefore for $|a|$, as a function of either the drive amplitude g or the driving frequency Ω. We shall analyze the dependence of the magnitude of the response on frequency in some detail, and leave it to the reader to perform a similar analysis of the dependence on drive amplitude, which is very similar.

In order to analyze the magnitude of the response $|a|$ as a function of driving frequency Ω let us differentiate the response function (1.30a),

$$\left[\frac{3}{64}\left(9+\eta^2\right)|a|^4+\frac{1}{4}(\eta-6\Omega)|a|^2+\frac{1}{4}+\Omega^2\right]d|a|^2 = \left[\frac{3}{4}|a|^4-2\Omega|a|^2\right]d\Omega. \quad (1.32)$$

This allows us immediately to find the condition for resonance, where the magnitude of the response is at its peak, by requiring that $d|a|^2/d\Omega=0$. We find that the resonance frequency Ω_{max} depends quadratically on the peak magnitude $|a|_{max}$, according to

$$\Omega_{max}=\frac{3}{8}|a|^2_{max}, \quad (1.33a)$$

or in terms of the original variables as

$$\tilde{\omega}_{max}=\omega_0+\frac{3}{8}\frac{\alpha}{m\omega_0}(\tilde{x}_0)^2_{max}. \quad (1.33b)$$

The curve, satisfying Eq. (1.33a), for which $|a|=\sqrt{8\Omega/3}$, is plotted in Fig. 1.3. It forms a square-root back-bone that connects all the resonance peaks for the different driving amplitudes, which is often seen in typical experiments with nanomechanical resonators. Thus the peak of the response is pulled further toward higher frequencies as the driving amplitude g is increased, as expected from a stiffening nonlinearity.

When the drive amplitude g is sufficiently strong, we can use Eq. (1.32) to find the two saddle-node bifurcation points, where the number of solutions changes from one to three and then back from three to one. At these points $d\Omega/d|a|^2=0$, yielding a quadratic equation in Ω whose solutions are

$$\Omega^{\pm}_{SN}=\frac{3}{4}|a|^2\pm\frac{1}{2}\sqrt{\frac{3}{16}\left(3-\eta^2\right)|a|^4-\eta|a|^2-1}. \quad (1.34)$$

When the two solutions are real, corresponding to the two bifurcation points, a linear stability analysis shows that the upper and lower branches of the response are stable solutions and the middle branch that exists for $\Omega_{SN}^{-} < \Omega < \Omega_{SN}^{+}$ is unstable. When the drive amplitude g is reduced, it approaches a critical value g_c where the two bifurcation points merge into an inflection point. At this point both $d\Omega/d|a|^2 = 0$ and $d^2\Omega/(d|a|^2)^2 = 0$, providing two equations for determining the critical condition for the onset of bistability, or the existence of two stable solution branches,

$$|a|_c^2 = \frac{8}{3}\frac{1}{\sqrt{3}-\eta}, \qquad \Omega_c = \frac{1}{2\sqrt{3}}\frac{3\sqrt{3}+\eta}{\sqrt{3}-\eta}, \qquad g_c{}^2 = \frac{32}{27}\frac{9+\eta^2}{\left(\sqrt{3}-\eta\right)^3}. \tag{1.35}$$

For the case without nonlinear damping, $\eta = 0$, the critical values are $|a|_c^2 = (4/3)^{3/2}$ and $\Omega_c = (3/4)^{1/2}$, for which the critical drive amplitude is $g_c = (4/3)^{5/4}$. For $0 < \eta < \sqrt{3}$, the critical driving amplitude g_c that is required for having bistability increases with η, as shown in Fig. 1.4. For $\eta > \sqrt{3}$ the discriminant in Eq. (1.34) is always negative, prohibiting the existence of bistability of solutions.

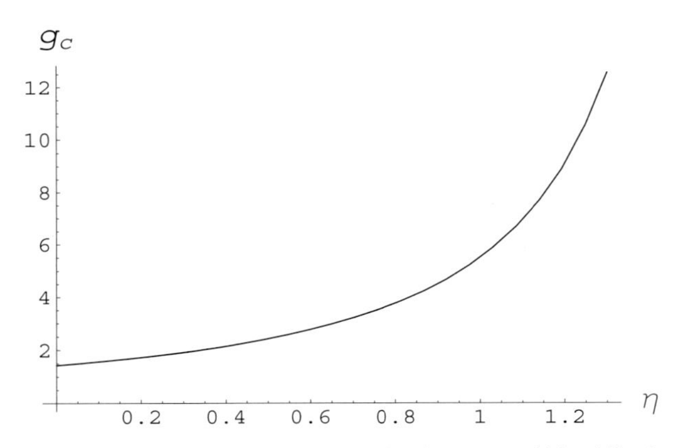

Fig. 1.4 Critical driving amplitude g_c for the onset of bistability in the response of the Duffing resonator as a function of nonlinear damping η, as given by Eq. (1.35).
Note that $g_c \to (4/3)^{(5/4)} \simeq 1.43$, as $\eta \to 0$.

Nonlinear damping acts to decrease the magnitude of the response when it is appreciable, that is when the drive amplitude is large. It gives rise to an effective damping rate for oscillations with magnitude $|a|$ that is given by $1 + \frac{1}{4}\eta|a|^2$, or in terms of the physical parameters by $\Gamma + \frac{1}{4}\tilde{\eta}\tilde{x}_0^2$. When viewing the response as it is plotted in Fig. 1.3, it is difficult to distinguish between the effects of the two forms of damping. The resonance peaks lie on the same back-bone regardless of the existence of a contribution from nonlinear damping. A more useful scheme for seeing the effect of nonlinear damping is to plot the response amplitude scaled by the drive $|a|/g$, often called the responsivity of the resonator, as shown in Fig. 1.5. Without nonlinear damping all peaks have the same height of 1. With nonlinear damping one clearly sees the decrease in the responsivity as the driving amplitude is increased.

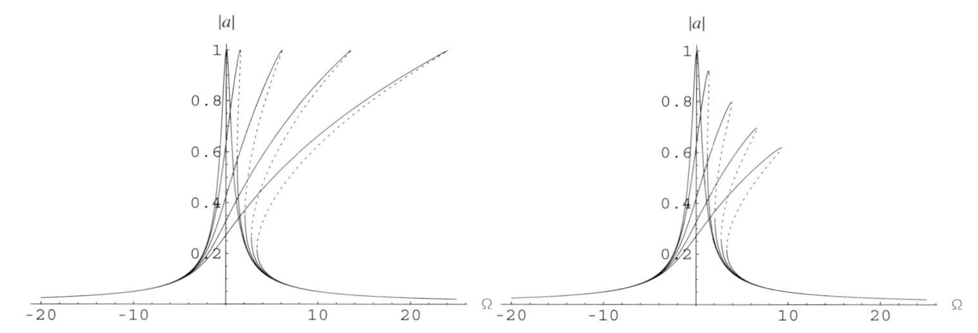

Fig. 1.5 Responsivity $|a|/g$ of the Duffing resonator without nonlinear damping (left) and with a small amount of nonlinear damping $\eta = 0.1$ (right) for different values of the driving amplitude g. Viewing the response in this way suggests an experimental scheme by which one could determine the importance of nonlinear damping and extract its magnitude.

The region of bistability, between the two saddle-node bifurcations (1.34) in the response of the driven Duffing resonator, is a source of a number of interesting dynamical features that are often observed in experiments with MEMS & NEMS resonators [3, 47–49]. Most obvious is the existence of hysteresis in quasistatic sweeps of either driving frequency or driving amplitude, which is readily observed in experiments. For example, if we start below resonance and sweep the frequency upwards along one of the constant drive-amplitude curves shown in Fig. 1.3, the response will gradually increase climbing up on the curve until it reaches the upper saddle-node bifurcation $\Omega^{+}_{SN}(g)$ when it will abruptly drop down to the lower stable solution branch, and continue toward lower response amplitudes to the right of the resonance. Upon switching the direction of the quasistatic sweep, the response amplitude will gradually increase until it reaches the lower saddle-node bifurcation $\Omega^{-}_{SN}(g)$ where it will abruptly jump up to the upper stable solution branch, from there on gradually following it downwards towards lower frequencies with diminishing response amplitude.

Another interesting aspect involves the notion of basins of attraction. If we fix the values of the driving amplitude and frequency, then depending on its initial conditions, the driven damped Duffing resonator will deterministically approach one of the two possible solutions. One can then map out the regions of the phase-space of initial conditions into the two so-called basins of attraction of the two possible stable solutions, where the unstable solution lies along the separatrix – the border line between the two basins of attraction. These basins of attraction were mapped out in a recent experiment using a suspended platinum nanowire by Kozinsky et al. [50]. If one additionally considers the existence of a random noise, which is always the case in real systems, then the separatrix becomes fuzzy, and it is possible to observe thermally activated switching of the resonator between its two possible solutions. What in fact is observed, say in an upward frequency scan, is that the resonator can drop to the small amplitude solution before it actually reaches the upper saddle-node bifurcation $\Omega^+_{SN}(g)$, and similarly for the lower bifurcation point. As the noise increases the observed size of the bistability region thus effectively shrinks. This was demonstrated with a doubly-clamped nanomechanical resonator made of aluminum nitride in a recent experiment by Aldridge and Cleland [21]. The existence of the saddle-node bifurcation has also been exploited for applications because at the bifurcation point the response of the resonator can change very dramatically if one changes the drive frequency or any of the resonator's physical parameters that can alter the response curve. This idea has been suggested and also used for signal amplification [26] as well as squeezing of noise [42,49].

Finally, much effort is invested these days to push experiments with nanomechanical resonators towards the quantum regime. In this context, it has been shown that the bistability region in the response of the driven damped Duffing resonator offers a novel approach for observing the transition from classical to quantum mechanical behavior as the temperature is lowered [51]. The essential idea is that one can find a regime in frequency and temperature where thermal switching between the two basins of attraction is essentially suppressed when the dynamics is classical, whereas if the resonator has already started entering the quantum regime, quantum dynamics allow it to switch between the two basins. Thus, an observation of switching can be used to ascertain whether or not a Duffing resonator is behaving quantum mechanically.

1.2.3 Addition of Other Nonlinear Terms

It is worth taking some time to consider the addition of other nonlinear terms that were not included in our original equation of motion (1.18). Without increasing the order of the nonlinearity, we could still add quadratic symmetry-breaking terms of the form x^2, $x\dot{x}$, and $\dot{x}^2$, as well as additional cubic damping terms of the form $\dot{x}^3$, and $x\dot{x}^2$. Such terms may appear naturally in actual physical situations, like the examples discussed in §1.1.2. For the reader who is anxious to skip on to the next section on parametrically-driven Duffing resonators, we state at the outset that the addition of such terms does not alter the response curves that we described in the previous section in any fundamental way. They merely conspire to renormalize the effective values of the coefficients used in the original equation of motion. Thus, without any particular model at hand, it is difficult to discern the existence of such terms in the equation.

Consider then an equation like (1.18), but with all these additional terms

$$m\frac{d^2\tilde{x}}{d\tilde{t}^2} + \Gamma\frac{d\tilde{x}}{d\tilde{t}} + m\omega_0^2\tilde{x} + \tilde{\beta}\tilde{x}^2 + \tilde{\mu}\tilde{x}\frac{d\tilde{x}}{d\tilde{t}} + \tilde{\rho}\left(\frac{d\tilde{x}}{d\tilde{t}}\right)^2 + \tilde{\alpha}\tilde{x}^3 + \tilde{\eta}\tilde{x}^2\frac{d\tilde{x}}{d\tilde{t}} + \tilde{\nu}\tilde{x}\left(\frac{d\tilde{x}}{d\tilde{t}}\right)^2 + \tilde{\zeta}\left(\frac{d\tilde{x}}{d\tilde{t}}\right)^3 = \tilde{G}\cos\tilde{\omega}\tilde{t}, \quad (1.36)$$

and perform the same scaling as in Eq. (1.20) for the additional parameters,

$$\beta = \frac{\tilde{\beta}}{\omega_0\sqrt{m\tilde{\alpha}}}; \quad \mu = \frac{\tilde{\mu}}{\sqrt{m\tilde{\alpha}}}; \quad \rho = \frac{\tilde{\rho}\omega_0}{\sqrt{m\tilde{\alpha}}}; \quad \nu = \frac{\tilde{\nu}\omega_0^2}{\tilde{\alpha}}; \quad \zeta = \frac{\tilde{\zeta}\omega_0^3}{\tilde{\alpha}}. \quad (1.37)$$

This yields, after performing the same scaling as before with $\epsilon = Q^{-1}$, a scaled equation of motion with all the additional nonlinearities,

$$\ddot{x} + \epsilon\dot{x} + x + \beta x^2 + \mu x\dot{x} + \rho\dot{x}^2 + x^3 + \eta x^2\dot{x} + \nu x\dot{x}^2 + \zeta\dot{x}^3 = \epsilon^{3/2}g\cos\omega t, \quad (1.38)$$

The important difference between this equation and the one we solved earlier (1.21) is that with the same scaling of x with $\sqrt{\epsilon}$ as before we now have terms on the order of ϵ. We therefore need to slightly change our trial expansion to contain such terms as well,

$$x(t) = \sqrt{\epsilon}x_0(t,T) + \epsilon x_{1/2}(t,T) + \epsilon^{3/2}x_1(t,T) + \ldots, \quad (1.39)$$

with $x_0 = \frac{1}{2}\left[A(T)e^{it} + c.c.\right]$ as before.

We begin by collecting all terms on the order of ϵ, arriving at

$$\ddot{x}_{1/2} + x_{1/2} = -\frac{1}{2}(\beta+\rho)|A|^2 - \frac{1}{4}\left[(\beta - \rho + i\mu)A^2e^{2it} + c.c.\right]. \quad (1.40)$$

This equation for the first correction $x_{1/2}(t)$ contains no secular terms and therefore can be solved immediately to give

$$x_{1/2}(t) = -\frac{1}{2}(\beta+\rho)|A|^2 + \frac{1}{12}\left[(\beta - \rho + i\mu)A^2e^{2it} + c.c.\right]. \quad (1.41)$$

We substitute this solution into the ansatz (1.39) and back into the equation of motion (1.38), and proceed to collecting terms on the order of $\epsilon^{3/2}$. We find a number of additional terms at this order that did not appear earlier on the right-hand side of Eq. (1.25) for the correction $x_1(t)$,

$$
\begin{aligned}
&-2\beta x_0 x_{1/2} - \mu\left(x_0 \dot{x}_{1/2} + \dot{x}_0 x_{1/2}\right) - 2\rho \dot{x}_0 \dot{x}_{1/2} - \nu x_0 \dot{x}_0^2 - \zeta \dot{x}_0^3 \\
&= \left\{\left[\frac{5}{12}\beta(\beta+\rho) + \frac{1}{6}\rho^2 + \frac{1}{24}\mu^2 - \frac{1}{8}\nu\right] + i\left[\frac{1}{8}\mu(\beta+\rho) - \frac{3}{8}\zeta\right]\right\}|A|^2 A e^{it} \\
&\quad + \text{nonsecular terms.}
\end{aligned} \tag{1.42}
$$

After adding the additional secular terms, we obtain a modified equation for the slowly varying amplitude $A(T)$,

$$
\begin{aligned}
\frac{dA}{dT} &= -\frac{1}{2}A + i\frac{3}{8}\left(1 - \frac{10}{9}\beta(\beta+\rho) - \frac{4}{9}\rho^2 - \frac{1}{9}\mu^2 + \frac{1}{3}\nu\right)|A|^2 A \\
&\quad - \frac{1}{8}\left(\eta - \mu(\beta+\rho) + 3\zeta\right)|A|^2 A - i\frac{g}{2}e^{i\Omega T} \\
&\equiv -\frac{1}{2}A + i\frac{3}{8}\alpha_{eff}|A|^2 A - \frac{1}{8}\eta_{eff}|A|^2 A - i\frac{g}{2}e^{i\Omega T}.
\end{aligned} \tag{1.43}
$$

We find that the equation is formally identical to the one we had earlier (1.26) before adding the extra nonlinear terms. The response curves and the discussion of the previous section therefore still apply after taking into account all the quadratic and cubic nonlinear terms. All these terms combine together in a particular way giving rise to the two effective cubic parameters, defined in (1.43). This, in fact, allows one some flexibility in tuning the nonlinearities of a Duffing resonator in real experimental situations. For example, Kozinsky *et al.* [52] use this flexibility to tune the effective Duffing parameter α_{eff} using an external electrostatic potential, as described in § 1.1.3 and shown in Fig. 1.1, thus affecting both the quadratic parameter $\tilde{\beta}$ and the cubic parameter $\tilde{\alpha}$ in the physical equation of motion (1.36). Note that due to the different signs of the various contributions to the effective nonlinear parameters, one could actually cause the cubic terms to vanish, altering the response in a fundamental way.

1.3
Parametric Excitation of a Damped Duffing Resonator

Parametric excitation offers an alternative approach for actuating MEMS or NEMS resonators. Instead of applying an external force that acts directly on the resonator, one modulates one (or more) of its physical parameters as a function of time, which in turn modulates the normal frequency of the resonator. This is what happens on a swing, where the up-and-down motion of the center of mass of the swinging child effectively changes the length of the swing, thereby modulating its natural frequency. The most effective way to swing is to move the center of mass up-and-down twice in every period of oscillation, but one can also swing by moving up-and-down at slower rates, namely once every n^{th} multiple of half a period, for any integer n. Let H be the relative amplitude by which the normal frequency is modulated, and ω_P be the frequency of the modulation, often called the pump frequency. One can show [53] that there is a sequence of tongue shaped regions in the $H - \omega_P$ plane where the smallest fluctuations away from the quiescent state of the swing, or any other parametrically-excited resonator [15], are exponentially amplified. This happens when the amplitude of the modulation H is sufficiently strong to overcome the effect of damping. The threshold for the nth instability tongue scales as $(Q^{-1})^{1/n}$. Above this threshold, the amplitude of the motion grows until it is saturated by nonlinear effects. We shall describe the nature of these oscillations for driving above threshold later on, both for the first ($n = 1$) and the second ($n = 2$) instability tongues, but first we shall consider the dynamics when the driving amplitude is just below threshold, as it also offers interesting behavior, and a possibility for novel applications such as parametric amplification [23, 24, 54] and noise squeezing [23].

There are a number of actual schemes for the realization of parametric excitation in MEMS & NEMS devices. The simplest, and probably the one most commonly used in the micron scale, is to use an external electrode that can induce an external potential that, if modulated in time, changes the effective spring constant of the resonator [15, 18, 19, 31, 55, 56]. Based on our treatment of this situation in § 1.1.3, this method is likely to modulate all the coefficients in the potential felt by the resonator, thus modulating, for example, also the Duffing parameter α. Similarly, one may devise configurations in which an external electrode deflects a doubly-clamped beam from its equilibrium, thereby inducing extra tension within the beam itself, as described in § 1.1.4, that can be modulated in time. Alternatively, one may generate motion in the clamps holding a doubly-clamped beam by its ends, as shown for example in Fig. 1.6, thus inducing in it a time-varying tension, which is likely to affect the other physical parameters to a lesser extent. These methods allow one to modulate the tension in the beam directly and thus modulate its normal frequency. More recently, Masmanidis et al. [57] developed layered piezoelectric NEMS structures whose tension can be fine tuned in doubly-clamped configurations, thus enabling fine control of the normal frequency of the beam with a simple turn of a knob.

Only a minor change is required in our equation of the driven damped Duffing resonator to accommodate this new situation, namely the addition of a modulation of the linear spring constant. Beginning with the scaled form of

Fig. 1.6 A configuration that uses electromotive actuation to perform parametric excitation of a doubly clamped beam – the central segment of the H-shpaed device. A static magnetic field runs nornal to the plane of the device. A metallic wire that runs along the external suspended segments of the H-device carries alternating current in opposite directions, thus applying opposite Lorentz forces that induce a time-varying compression of the central segment. This modulates the tension in the central segment, thus varying its normal frequency. This configuration was recently used by Karabalin et al. to demonstrate parametric amplification of a signal running along the central beam through a separate electric circuit. Image courtesy of Michael Roukes.

the Duffing equation (1.19) we obtain

$$\ddot{x} + Q^{-1}\dot{x} + [1 + H\cos\omega_P t]\,x + x^3 + \eta x^2 \dot{x} = G\cos(\omega_D t + \phi_g)\,, \tag{1.44}$$

where the scaling is the same as before, and we shall again use the damping Q^{-1} to define the small expansion parameter ϵ. The term proportional to H on the left hand side is the external drive that modulates the spring constant, giving a term that is proportional to the displacement x as well as to the strength of the drive – this is the parametric drive. We consider first the largest excitation effect that occurs when the pump frequency is close to twice the resonant frequency of the resonator. This is the region in the $H - \omega_P$ plane that we termed the first instability tongue. We therefore take the pump frequency to be an amount $\epsilon\Omega_P$ away from twice the resonant frequency, and take the drive amplitude to scale as the damping, *i.e.* we set $H = \epsilon h$. The term on the right hand side is a direct additive drive or signal, with amplitude scaled as in the discussion of the Duffing equation. The frequency of the drive is an amount $\varepsilon\Omega_D$ away from the resonator frequency which has been scaled to 1.

The scaled equation of motion that we now treat in detail is therefore

$$\ddot{x} + \epsilon\dot{x} + \Big(1 + \epsilon h \cos\big[(2 + \epsilon\Omega_P)\, t\big]\Big) x + x^3 + \eta x^2 \dot{x}$$
$$= \epsilon^{3/2}\, |g| \cos\big[(1 + \epsilon\Omega_D)t + \phi_g\big], \quad (1.45)$$

where we now use $g = |g|\, e^{i\phi_g}$ to denote a complex drive amplitude. We follow the same scheme of secular perturbation theory as in §1.2.2, using a trial solution in the form of Eq. (1.22) and proceeding as before. The new secular term, appearing on the right-hand side of Eq. (1.25), arising from the parametric drive is

$$-\frac{1}{4} h A^* e^{i\Omega_P T} e^{it}. \quad (1.46)$$

This gives the equation for the slowly varying amplitude,

$$\frac{dA}{dT} + \frac{1}{2}A - i\frac{h}{4}A^* e^{i\Omega_P T} - i\frac{3}{8}|A|^2 A + \frac{\eta}{8}|A|^2 A = -i\frac{g}{2}e^{i\Omega_D T}. \quad (1.47)$$

1.3.1
Driving Below Threshold: Amplification and Noise Squeezing

We first study the amplitude of the response of a parametrically-pumped Duffing resonator to an external direct drive $g \neq 0$. We will see that the characteristic behavior changes from amplification of an applied signal to oscillations, even in the absence of a signal, at a critical value of $h = h_c = 2$. It is therefore convenient to introduce a reduced parametric drive $\bar{h} = h/h_c = h/2$ which plays the role of a bifurcation parameter with a critical value of 1. We begin by assuming that the drive is small enough so that the magnitude of the response remains small and the nonlinear terms in Eq. (1.47) can be neglected. This gives the linear equation

$$\frac{dA}{dT} + \frac{1}{2}A - i\frac{\bar{h}}{2}A^* e^{i\Omega_P T} = -i\frac{g}{2}e^{i\Omega_D T}. \quad (1.48)$$

In general, at long times after transients have died out, the solution will take the form

$$A = a' e^{i\Omega_D T} + b' e^{i(\Omega_P - \Omega_D)T} \quad (1.49)$$

with a', b' complex constants.

We first consider the degenerate case where the pump frequency is tuned to be always twice the signal frequency. In this case $\Omega_P = 2\Omega_D$ and the long time solution is

$$A = ae^{i\Omega_D T} \tag{1.50}$$

with a a time independent complex amplitude. Substituting this into Eq. (1.48) gives

$$(2\Omega_D - i)a - \bar{h}a^* = -g. \tag{1.51}$$

Equation (1.51) is easily solved. If we first look on resonance, $\Omega_D = 0$ we find

$$a = e^{i\pi/4}\left[\frac{\cos(\phi_g + \pi/4)}{(1-\bar{h})} + i\frac{\sin(\phi_g + \pi/4)}{(1+\bar{h})}\right]|g| \tag{1.52}$$

where we remind the reader that $g = |g|e^{i\phi_g}$ so that ϕ_g measures the phase of the signal relative to the pump. Equation (1.52) shows that on resonance and for $\bar{h} \to 1$ (or $h \to h_c = 2$) there is strongest *enhancement* of the response for a signal that has a phase $-\pi/4$ relative to the pump. Physically this means that the maximum of the signal occurs a quarter of a pump cycle *after* a maximum of the pump. (The phase $3\pi/4$ gives the same result: this corresponds to shifting the oscillations by a complete pump period.) The enhancement diverges as $\bar{h} \to 1$, *provided* that the signal amplitude g is small enough that the enhanced response remains within the linear regime. For a fixed signal amplitude g, the response will become large as $\bar{h} \to 1$ so that the nonlinear terms in Eq. (1.47) must be retained, and the expressions we have derived no longer hold. This situation is discussed in the next section.

On the other hand, for a signal that has a relative phase $\pi/4$ or $5\pi/4$ (the maximum of the signal occurring a quarter of a pump cycle *before* a maximum of the pump) there is a weak suppression, by a factor of 2 as $\bar{h} \to 1$. A noise signal on the right-hand side of the equation of motion (1.45) would have both phase components. This leads to the *squeezing* of the noisy displacement driven by this noise, with the response at phase $-\pi/4$ amplified and the response at phase $\pi/4$ quenched.

The full expression for $\Omega_D \neq 0$ for the amplitude of the response is

$$a = -\left[\frac{2\Omega_D + (i + \bar{h}e^{-2i\phi_g})}{4\Omega_D^2 + (1 - \bar{h}^2)}\right] g. \tag{1.53}$$

For $\bar{h} \to 1$ the response is large when $\Omega_D \ll 1$, that is for frequencies much closer to resonance than the original width of the resonator response. In these limits the first term in the numerator may be neglected *unless* $\phi_g \simeq \pi/4$. This then gives

$$|a| = \frac{2\left|g\cos(\phi_g + \pi/4)\right|}{4\Omega_D^2 + (1-\bar{h}^2)}. \tag{1.54}$$

This it not the same as the expression for a resonant response, since the frequency dependence of the amplitude, not amplitude squared, is Lorentzian. However, estimating a quality factor from the width of the sharp peak would give an *enhanced* quality factor $\propto 1/\sqrt{1-\bar{h}^2}$, becoming very large as $\bar{h} \to 1$. For the case $\phi_g = \pi/4$ the magnitude of the response is

$$\left|a_{\phi_g=\pi/4}\right| = -\frac{\sqrt{4\Omega_D^2 + (1-\bar{h})^2}}{4\Omega_D^2 + (1-\bar{h}^2)}\left|\bar{g}\right|. \tag{1.55}$$

This initially increases as the frequency approaches resonance, but decreases for $\Omega_D \lesssim \sqrt{1-\bar{h}}$, approaching $|g|/2$ for $\Omega_D \to 0, \bar{h} \to 1$.

For the general or nondegenerate case of $\Omega_P \neq 2\Omega_D$ it is straightforward to repeat the calculation with the ansatz Eq. (1.49). The result is

$$a' = -\frac{2(\Omega_P - \Omega_D) + i}{4\Omega_D(\Omega_P - \Omega_D) - 2i(\Omega_P - 2\Omega_D) + 1 - \bar{h}^2}g. \tag{1.56}$$

Notice that this does *not* reduce to Eq. (1.53) for $\Omega_P = 2\Omega_D$, since we miss some of the interference terms in the degenerate case if we base the calculation on $\Omega_P \neq 2\Omega_D$. Also, of course, there is no dependence of the magnitude of the response on the phase of the signal ϕ_g, since for different frequencies the phase difference cannot be defined independent of an arbitrary choice of the origin of time. If the pump frequency is maintained fixed at twice the resonator resonance frequency, corresponding to $\Omega_P = 0$, the expression for the amplitude of the response simplifies to

$$a' = \frac{2\Omega_D - i}{-4\Omega_D^2 + 4i\Omega_D + 1 - \bar{h}^2}g. \tag{1.57}$$

Again there is an enhanced response for drive frequencies closer to resonance than the width of the original resonator response. In this region $\Omega_D \ll 1$ so that

$$\left|a'\right| \simeq |g|\frac{1}{\sqrt{(4\Omega_D)^2 + (1-\bar{h}^2)^2}}. \tag{1.58}$$

This is the usual Lorentzian describing a resonance with a quality factor enhanced by $(1-\bar{h}^2)^{-1}$, as shown on the right-hand side of Fig. 1.7.

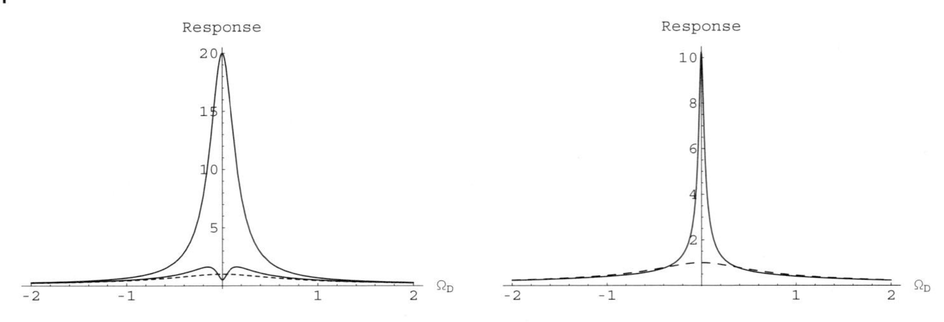

Fig. 1.7 Response of the parametrically driven resonator as the signal frequency Ω_D varies for pump frequency equal to twice the signal frequency (left) and for the pump frequency fixed at the linear resonance frequency (right), given by Eqs. (1.53) and (1.57), respectively. The dashed curve is the response of the resonator to the same signal without parametric drive. In the left hand figure the upper curve is for the amplified phase $\phi_g = -\pi/4$, and the lower curve is for the phase $\phi_g = \pi/4$, giving squeezing on resonance. In both cases the reduced pump amplitude $\bar{h} = h/h_c$ is 0.95.

For the resonance condition $\Omega_D = \Omega_P = 0$, corresponding to a pump frequency twice the resonance frequency of the device, and a signal at this resonant frequency, the response amplitude in the linear approximation diverges as the pump amplitude approaches the critical value $h_c = 2$. This is the signature of a linear instability to self sustained oscillations in the absence of any drive. We analyze this *parametric instability* in the next section.

1.3.2
Linear Instability

The divergence of the response as $\bar{h}$ approaches unity from below corresponding to $h \to 2$, suggests a linear instability for $h > 2$, or $QH > 2$ in the original units. We can see this directly from Eq. (1.47) setting $g = 0$ but still ignoring the nonlinear terms, to give the linear equation

$$\frac{dA}{dT} + \frac{1}{2}A = i\frac{h}{4}A^* e^{i\Omega_P T}. \tag{1.59}$$

We seek a solution of the form

$$A = |a|\, e^{i\phi} e^{\sigma T} e^{i(\Omega_P/2)T} \tag{1.60}$$

with a real σ giving exponential growth or decay. Substituting into Eq. (1.59) gives

$$\sigma = \frac{-1 \pm \sqrt{(h/2)^2 - \Omega_P^2}}{2}, \tag{1.61}$$

$$\phi = \pm\left[\frac{\pi}{4} - \frac{1}{2}\arcsin\left(\frac{2\Omega_P}{h}\right)\right] \tag{1.62}$$

where we take the value of arcsin between 0 and $\pi/2$ and the plus and minus signs in the two equations go together. Note that these expressions apply for $h/2 > \Omega_P$; for $h/2 < \Omega_P$ the value of σ is complex. For pumping at twice the resonance frequency $\Omega_P = 0$, one phase of oscillation $\phi = \frac{\pi}{4}$ has a *reduced damping*, with $\sigma = -(1/2 - h/4)$ for $h < 2$, and an *instability*, $\sigma = (h/4 - 1/2) > 0$ signaling exponential growth, for $h > 2$. The other phase of oscillation $\phi = -\frac{\pi}{4}$ has an *increased damping*, with $\sigma = -(1/2 + h/4)$. The general condition for instability is

$$h > 2\sqrt{1 + \Omega_P^2}, \tag{1.63}$$

showing an increase of the threshold for nonzero frequency detuning Ω_P, as shown in Fig. 1.8. The linear instability that occurs for positive σ gives exponentially growing solutions that eventually saturate due to nonlinearity.

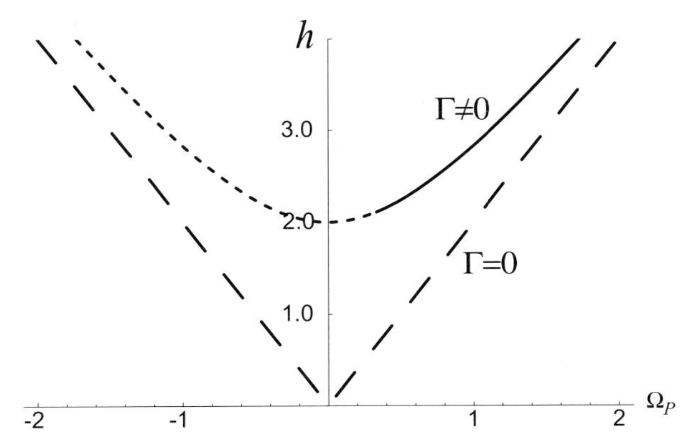

Fig. 1.8 The first instability tongue of the parametrically-driven Duffing resonator – the threshold for instability – plotted in the (Ω_P, h) plane. The lower, long-dashed curve shows the threshold without any linear damping ($\Gamma = 0$), which is zero on resonance. The upper curve shows the threshold with linear damping ($\Gamma \neq 0$). The threshold on resonance ($\Omega_P = 0$) is $h = 2$. The solid and short-dashed regions of the upper curve indicate the so-called subcritical and supercritical branches of the instability, respectively, as discussed in Section 1.3.4. On the subcritical branch ($\Omega_P > 4\eta/3$) there will be hysteresis as h is varied, and on the supercritical branch ($\Omega_P < 4\eta/3$) there will not be any hysteresis.

1.3.3 Nonlinear Behavior Near Threshold

Nonlinear effects may also be important below the threshold of the parametric instability in the presence of a periodic signal or noise. As we have seen, in the linear approximation, the gain below threshold diverges in the linear approximation as $h \to h_c$. This is unphysical, and for a given signal or noise strength there is some h close enough to h_c where nonlinear saturation of the gain will become important. This will give a smooth behavior of the response of the driven system as h passes through h_c into the unstable regime. We first analyze the effects of nonlinearity near the threshold of the instability, and calculate the smooth behavior as h passes through h_c in the presence of an applied signal. In the following section we study the effects of nonlinearity on the self-sustained oscillations above threshold with more generality.

We take h to be close to h_c, and we take the signal to be small. This introduces a second level of "smallness". We have already assumed that the damping is small and that the deviation of the pump frequency from resonance is small. This means that the critical parametric drive H_c is also small. We now assume that $|H - H_c|$ is small compared with H_c, or equivalently in scaled units that $|h - h_c|$ is small compared with h_c, and introduce the perturbation parameter δ to implement this, *i.e.*, we assume that

$$\delta = \frac{h - h_c}{h_c} \ll 1. \tag{1.64}$$

We now use the same type of secular perturbation theory as the method leading to Eq. (1.47) to develop the expansion in δ. For simplicity we will develop the theory for the most interesting case of resonant pump and signal frequencies $\Omega_P = \Omega_D = 0$. The critical value of h is then $h_c = 2$, and the solution to Eq. (1.47) that becomes marginally stable at this value is

$$A = be^{i\pi/4}, \tag{1.65}$$

with b a real constant.

For h near h_c we make the ansatz for the solution

$$A = \delta^{1/2} b_0(\tau) e^{i\pi/4} + \delta^{3/2} b_1(\tau) + \cdots \tag{1.66}$$

with b_0 a real function of $\tau = \delta T$, a new, even slower, time scale that sets the scale for the time variation of the real amplitude b_0 near threshold. We must also assume the signal amplitude is very small, $g = \delta^{3/2}\hat{g}$, altogether having $G = (\epsilon\delta)^{3/2}\hat{g}$. Substituting Eq. (1.66) into Eq. (1.47) and collecting terms at $O(\delta^{3/2})$ yields

$$\frac{1}{2}(b_1 - b_1^*) = -\frac{\hat{g}}{2}e^{i\pi/4} - \frac{db_0}{d\tau} + \frac{1}{2}b_0 + i\frac{3}{8}b_0^3 - \frac{\eta}{8}b_0^3. \tag{1.67}$$

The left-hand side of this equation is necessarily imaginary, so in order to have a solution for b_1, so that the perturbation expansion is valid, the real part of the right-hand side must be zero. This is the solvability condition for the secular perturbation theory. This gives

$$\frac{db_0}{d\tau} = \frac{1}{2}b_0 - \frac{\eta}{8}b_0^3 - \frac{|\hat{g}|}{2}\cos(\phi_g + \pi/4). \tag{1.68}$$

It is more informative to write this equation in terms of the variables without the δ scaling. Introducing the "unscaled" amplitude $b = \delta^{1/2}b_0$, generalizing Eq. (1.65) so that

$$A = be^{i\pi/4} + O(\delta^{3/2}), \tag{1.69}$$

we can write the equation as

$$\frac{db}{dT} = \frac{1}{2}\frac{h - h_c}{h_c}b - \frac{\eta}{8}b^3 - \frac{|g|}{2}\cos(\phi_g + \pi/4). \tag{1.70}$$

Equation (1.70) can be used to investigate many phenomena, such as transients above threshold, and how the amplitude of the response to a signal varies as h passes through the instability threshold. The unphysical divergence of the response to a small signal as $h \to h_c$ from below is now eliminated. For example, exactly at threshold $h = h_c$ we have

$$|b| = \left(\frac{4}{\eta}\left|g\cos(\phi_g + \pi/4)\right|\right)^{1/3} \tag{1.71}$$

giving a finite response, but one proportional to $|g|^{1/3}$ rather than to $|g|$. The gain $|b/g|$ scales as $|g|^{-2/3}$ for $h = h_c$, and gets smaller as the signal gets larger, as shown in Fig. 1.9. Note that the physical origin of the saturation at the lowest order of perturbation theory is nonlinear damping. Without nonlinear damping the response amplitude (1.71) still diverges. With linear damping being also small, one would need to go to higher orders of perturbation theory to find a different physical mechanism that can provide this kind of saturation. The response to noise can also be investigated, by replacing the $|g|\cos(\phi_g + \pi/4)$ drive by a noise function. Equation (1.70), and the noisy version, appear in many contexts of phase transitions and bifurcations, and so solutions are readily found in Ref. [46].

1.3.4 Nonlinear Saturation Above Threshold

The linear instability leads to exponential growth of the amplitude, regardless of the signal, and we need to turn to the full nonlinear equation (1.47) with

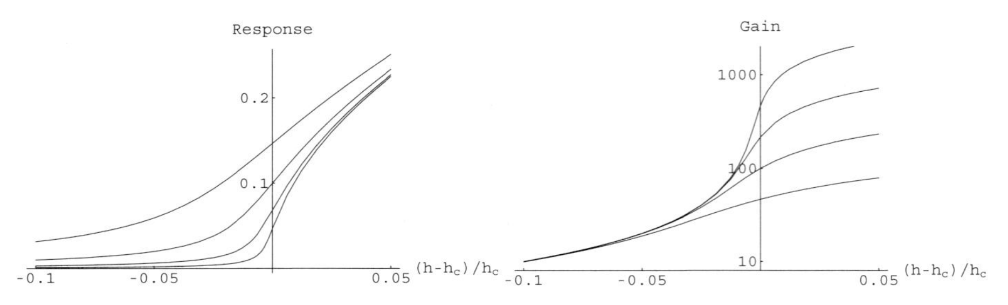

Fig. 1.9 Saturation of the response b (left) and gain $|b/g|$ (right) as the parametric drive h passes through the critical value h_c, for four different signal levels g. The signal levels are $\sqrt{\eta/4}$ times $10^{-2.5}$, 10^{-3}, $10^{-3.5}$, and 10^{-4}, increasing upwards for the response figure, and downwards for the gain figure. The response amplitude is also measured in units of $\sqrt{\eta/4}$. The phase of the signal is $\phi_g = -\pi/4$.

$g = 0$ to understand the saturation of this amplitude. Ignoring initial transients, and assuming that the nonlinear terms in the equation are sufficient to saturate the growth of the instability, we try a steady-state solution of the form

$$A(T) = ae^{i\left(\frac{\Omega_P}{2}\right)T}. \tag{1.72}$$

This amplitude a can be any solution of the equation

$$\left[\left(\frac{3}{4}|a|^2 - \Omega_P\right) + i\left(1 + \frac{\eta}{4}|a|^2\right)\right] a = \frac{h}{2}a^*, \tag{1.73}$$

obtained by substituting the steady-state solution (1.72) into the equation (1.47) of the secular terms. We immediately see that having no response ($a = 0$) is always a possible solution regardless of the excitation frequency Ω_P. Expressing $a = |a|e^{i\phi}$ we obtain, after taking the magnitude squared of both sides, the intensity $|a|^2$ of the non-trivial response as all positive roots of the equation

$$\left(\Omega_P - \frac{3}{4}|a|^2\right)^2 + \left(1 + \frac{\eta}{4}|a|^2\right)^2 = \frac{h^2}{4}. \tag{1.74}$$

After having noted that $|a| = 0$ is always a solution, we are left with a quadratic equation for $|a|^2$ and therefore, at most, two additional positive solutions for $|a|$. This has the form of a distorted ellipse in the $(\Omega_P, |a|^2)$ plane, and a parabola in the $(|a|^2, h)$ plane. In addition, we obtain for the relative phase of the response

$$\phi = \frac{i}{2}\ln\frac{a^*}{a} = -\frac{1}{2}\arctan\frac{1 + \frac{\eta}{4}|a|^2}{\frac{3}{4}|a|^2 - \Omega_P}. \tag{1.75}$$

In Fig. 1.10 we plot the response intensity $|a|^2$ of a Duffing resonator to parametric excitation as a function of the pump frequency Ω_P, for a fixed scaled drive amplitude $h = 3$. Solid curves indicate stable solutions, and dashed curves are solutions that are unstable to small perturbations. Thin curves show the response without nonlinear damping ($\eta = 0$) which grows indefinitely with frequency Ω_P and is therefore incompatible with experimental observations [15,19,58] and the assumptions of our calculation. Again, as we saw for the saturation below threshold, without nonlinear damping and with linear damping being small, one would have to go to higher orders of perturbation theory to search for a physical mechanism that could provide saturation. For large linear damping, or small Q, one sees saturation even without nonlinear damping [59]. Thick curves show the response with finite nonlinear damping ($\eta = 1$). With finite η there is a maximum value for the response $|a|^2_{max} = 2(h-2)/\eta$, and a maximum frequency,

$$\Omega_{SN} = \frac{h}{2}\sqrt{1+\left(\frac{3}{\eta}\right)^2} - \frac{3}{\eta}, \tag{1.76}$$

at a saddle-node bifurcation, where the stable and unstable nontrivial solutions meet. For frequencies above Ω_{SN} the only solution is the trivial one $a = 0$. These values are indicated by horizontal and vertical dotted lines in Fig. 1.10.

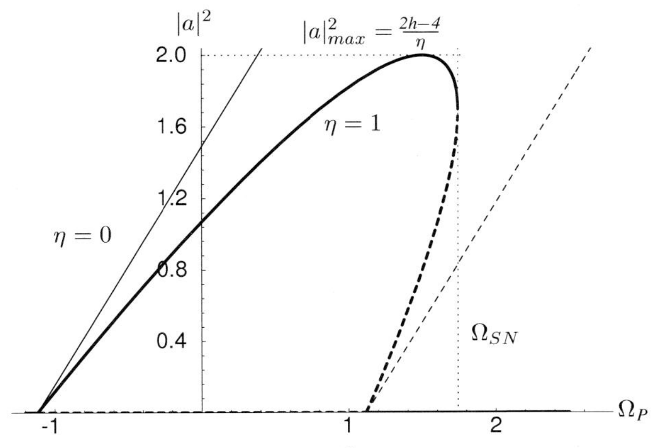

Fig. 1.10 Response intensity $|a|^2$ as a function of the pump frequency Ω_P, for fixed amplitude $h = 3$. Solid curves are stable solutions; dashed curves are unstable solutions. Thin curves show the response without non-linear damping ($\eta = 0$). Thick curves show the response for finite nonlinear damping ($\eta = 1$). Dotted lines indicate the maximal response intensity $|a|^2_{max}$ and the saddle-node frequency Ω_{SN}.

The threshold for the instability of the trivial solution is easily verified by setting $a = 0$ in the expression (1.74) for the nontrivial solution, or by inverting the expression (1.63) for the instability that we obtained in the previous section. As seen in Fig. 1.10, for a given h the threshold is situated at $\Omega_P = \pm\sqrt{(h/2)^2 - 1}$. This is the same result calculated in the previous section, where we plotted the threshold tongue in Fig. 1.8, in the (h, Ω_P) plane. Figure 1.10 is a cut going horizontally through that tongue at a constant drive amplitude $h = 3$.

Like the response of a forced Duffing resonator (1.29), the response of a parametrically excited Duffing resonator also exhibits hysteresis in quasistatic frequency scans. If the frequency Ω_P starts out at negative values and is increased gradually with a fixed amplitude h, the zero response will become unstable as the lower threshold is crossed at $-\sqrt{(h/2)^2 - 1}$, after which the response will gradually increase along the thick solid curve in Fig. 1.10, until Ω_P reaches Ω_{SN} and the response drops abruptly to zero. If the frequency is then decreased gradually, the response will remain zero until Ω_P reaches the upper instability threshold $+\sqrt{(h/2)^2 - 1}$, and the response will jump abruptly to the thick solid curve above, and then gradually decrease to zero along this curve.

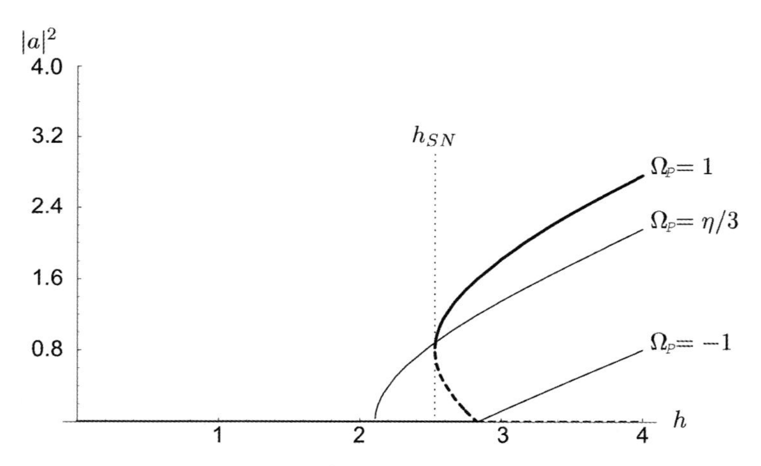

Fig. 1.11 Response intensity $|a|^2$ as a function of the parametric drive amplitude h for fixed frequency Ω_P and finite nonlinear damping ($\eta = 1$). Thick curves show the stable (solid curves) and unstable (dashed curves) response for $\Omega_P = 1$. Thin curves show the stable solutions for $\Omega_P = \eta/3$ and $\Omega_P = -1$, and demonstrate that hysteresis as h is varied is expected only for $\Omega_P > \eta/3$.

Finally, in Fig. 1.11 we plot the response intensity $|a|^2$ of the Duffing resonator as a function of drive amplitude h, for fixed frequency Ω_P and finite nonlinear damping $\eta = 1$. This would correspond to performing a verical cut through the instability tongue 1.8. Again, solid curves indicate stable solutions, and dashed curves unstable solutions. Thick curves show the response for $\Omega_P = 1$, and thin curves show the response for $\Omega_P = \eta/3$ and $\Omega_P = -1$. The intersection of the trivial and the nontrivial solutions, which corresponds to the instability threshold (1.63), occurs at $h = 2\sqrt{\Omega_P^2 + 1}$. For $\Omega_P < \eta/3$ the nontrivial solution for $|a|^2$ grows continuously for h above threshold and is stable. This is a supercritical bifurcation. On the other hand, for $\Omega_P > \eta/3$ the bifurcation is subcritical – the nontrivial solution grows for h *below* threshold. This solution is unstable until the curve of $|a|^2$ as a function of h bends around at a saddle-node bifurcation at

$$h_{SN} = \frac{2 + \frac{2\eta}{3}\Omega_P}{\sqrt{1 + \left(\frac{\eta}{3}\right)^2}}, \tag{1.77}$$

where the solution becomes stable and $|a|^2$ is once more an increasing function of h. For amplitudes $h < h_{SN}$ the only solution is the trivial one $a = 0$. Hysteretic behavior is therefore expected for quasistatic scans of the drive amplitude h only if the fixed frequency $\Omega_P > \eta/3$, as can be inferred from Fig. 1.11.

1.3.5
Parametric Excitation at the Second Instability Tongue

We wish to examine the second tongue – which is readily accessible in experiments because in this case the pump and the response frequencies are the same – by looking at the response above threshold and highlighting the main changes from the first tongue. We start with the general equation for a parametrically-driven Duffing resonator (1.44), but with no direct drive ($g = 0$), where the parametric excitation is performed around 1 and not around 2. Correspondingly, the scaling of H with respect to ϵ needs to be changed to $H = h\sqrt{\epsilon}$. The technical reason for doing this is that if we naively take $H = h\epsilon$ as before, the parametric driving term does not contribute to the order $\epsilon^{3/2}$ secular term which we use to find the response, and the order $\epsilon^{1/2}$ term in x becomes identically zero. Scaling H in this manner, as we shall immediately see, will introduce a non-secular correction to x at order ϵ, and this correction will contribute to the order $\epsilon^{3/2}$ secular term and will give us the required response. The equation of motion then becomes

$$\ddot{x} + x = -\frac{h\epsilon^{1/2}}{2}\left(e^{i(t+\Omega_P T)} + c.c.\right)x - \epsilon\dot{x} - x^3 - \eta x^2\dot{x}, \tag{1.78}$$

and we try an expansion of the solution of the form

$$x(t) = \epsilon^{1/2}\frac{1}{2}\left(A(T)e^{it} + c.c.\right) + \epsilon x_{1/2}(t) + \epsilon^{3/2}x_1(t) + \dots \tag{1.79}$$

Substituting this expansion into the equation of motion (1.78) we obtain at order $\epsilon^{1/2}$ the linear equation as usual, and at order ϵ

$$\ddot{x}_{1/2} + x_{1/2} = -\frac{h}{4}\left(Ae^{i\Omega_P T}e^{2it} + A^* e^{i\Omega_P T} + c.c.\right). \tag{1.80}$$

As expected, there is no secular term on the right-hand side, so we can immediately solve for $x_{1/2}$,

$$x_{1/2}(t) = \frac{h}{4}\left(\frac{A}{3}e^{i\Omega_P T}e^{2it} - A^* e^{i\Omega_P T} + c.c.\right) + O(\epsilon). \tag{1.81}$$

Substituting the solution for $x_{1/2}$ into the expansion (1.79), and the expansion back into the equation of motion (1.78), contributes an additional term from the parametric driving which has the form

$$\begin{aligned} &\epsilon^{3/2}\frac{h^2}{8}\left(-\frac{A}{3}e^{i\Omega_P T}e^{2it} + A^* e^{i\Omega_P T} + c.c.\right)\left(e^{i\Omega_P T}e^{it} + c.c.\right) \\ &\quad = \epsilon^{3/2}\frac{h^2}{8}\left(\frac{2}{3}A + A^* e^{i2\Omega_P T}\right)e^{it} + c.c. + \text{ non secular terms.} \end{aligned} \tag{1.82}$$

This gives us the required contribution to the equation for the vanishing secular terms. All other terms remain as they were in Eq. (1.47), so that the new equation for determining $A(T)$ becomes

$$\frac{dA}{dT} + i\frac{h^2}{8}\left(\frac{2}{3}A + A^* e^{i2\Omega_P T}\right) + \frac{1}{2}A - i\frac{3}{8}|A|^2 A + \frac{\eta}{8}|A|^2 A = 0. \tag{1.83}$$

Again, ignoring initial transients, and assuming that the nonlinear terms in the equation are sufficient to saturate the growth of the instability, we try a steady-state solution, this time of the form

$$A(T) = ae^{i\Omega_P T}. \tag{1.84}$$

The solution to the equation of motion (1.78) is therefore

$$x(t) = \epsilon^{1/2}(ae^{i(1+\epsilon\Omega_P)t} + c.c.) + O(\epsilon), \tag{1.85}$$

where the correction $x_{1/2}$ of order ϵ is given in Eq. (1.81) and, as before, we are not interested in the correction $x_1(t)$ of order $\epsilon^{3/2}$, but rather in the fixed amplitude a of the lowest order term. We substitute the steady-state solution (1.84) into the Eq. (1.83) of the secular terms and obtain

$$\left[\left(\frac{3}{4}|a|^2 - 2\Omega_P - \frac{h^2}{6}\right) + i\left(1 + \frac{\eta}{4}|a|^2\right)\right] a = \frac{h^2}{4}a^*. \tag{1.86}$$

By taking the magnitude squared of both sides we obtain, in addition to the trivial solution $a = 0$, a non-trivial response given by

$$\left(\frac{3}{4}|a|^2 - 2\Omega_P - \frac{1}{6}h^2\right)^2 + \left(1 + \frac{\eta}{4}|a|^2\right)^2 = \frac{h^4}{16}. \tag{1.87}$$

Figure 1.12 shows the response intensity $|a|^2$ as a function of the frequency Ω_P, for a fixed drive amplitude of $h = 3$ producing a horizontal cut through the second instability tongue. The solution looks very similar to the response shown in Fig. 1.10 for the first instability tongue, though we should point out two important differences. The first is that the orientation of the ellipse, indicated by the slope of the curves for $\eta = 0$, is different. The slope here is 8/3 whereas for the first instability tongue the slope is 4/3. The second is the change in the scaling of h with ϵ, or the inverse quality factor Q^{-1}. The lowest critical drive amplitude for an instability at the second tongue is again on resonance ($\Omega_P = 0$), and its value is again $h = 2$, but now this implies that $H\sqrt{Q} = 2$, or that H scales the square root of the linear damping rate Γ. This is consistent with the well known result (see, for example, [53, § 3]) that the minimal amplitude for the instability of the nth tongue scales as $\Gamma^{1/n}$.

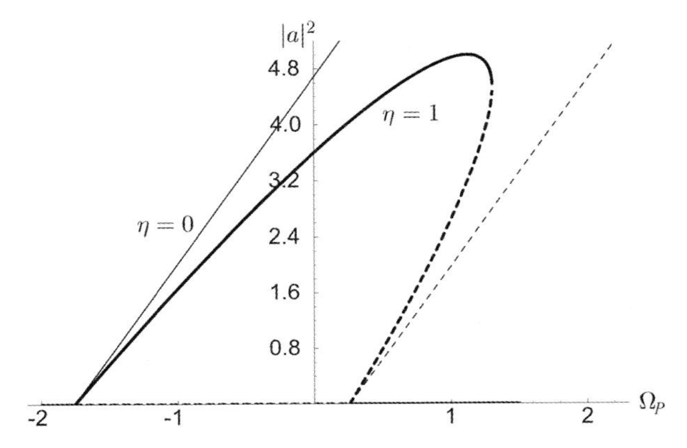

Fig. 1.12 Response intensity $|a|^2$ of a parametrically-driven Duffing resonator as a function of the pump frequency Ω_P, for a fixed amplitude $h = 3$ in the second instability tongue. Solid curves are stable solutions; dashed curves are unstable solutions. Thin curves show the response without non-linear damping ($\eta = 0$). Thick curves show the response for finite nonlinear damping ($\eta = 1$).

1.4
Parametric Excitation of Arrays of Coupled Duffing Resonators

The last two sections of this review describe theoretical work that was motivated directly by the experimental work of Buks and Roukes [37], who fabricated an array of nonlinear micromechanical doubly-clamped gold beams, and excited them parametrically by modulating the strength of an externally-controlled electrostatic coupling between neighboring beams. The Buks and Roukes experiment was modeled by Lifshitz and Cross [60, henceforth LC] using a set of coupled nonlinear equations of motion. They used secular perturbation theory – as we have described so far for a system with just a single degree of freedom – to convert these equations of motion into a set of coupled nonlinear *algebraic* equations for the normal mode amplitudes of the system, enabling them to obtain exact results for small arrays, but only a qualitative understanding of the dynamics of large arrays. We shall review these results in this section.

In order to obtain analytical results for large arrays, Bromberg, Cross, and Lifshitz [61, henceforth BCL] studied the same system of equations, approaching it from the continuous limit of infinitely-many degrees of freedom, and obtaining a description of the slow spatiotemporal dynamics of the array of resonators in terms of an amplitude equation. BCL showed that this amplitude equation could predict the initial mode that develops at the onset of parametric oscillations as the driving amplitude is gradually increased from zero, as well as a sequence of subsequent transitions to other single-mode oscillations. We shall review these results in Section 1.5. Kenig et al. [62] are currently extending the investigation of the amplitude equation to more general questions such as pattern selection when many patterns or solutions are simultaneously stable, as well as other experimentally-relevant question such as the response of the system of coupled resonators to time dependent sweeps of the control parameters, rather than quasistatic sweeps like the ones we have been discussing here.

1.4.1
Modeling an Array of Coupled Duffing Resonators

LC modeled the array of coupled nonlinear resonators that was studied by Buks and Roukes using a set of coupled equations of motion (EOM) of the form

$$\ddot{u}_n + u_n + u_n^3 - \frac{1}{2}Q^{-1}(\dot{u}_{n+1} - 2\dot{u}_n + \dot{u}_{n-1})$$
$$+ \frac{1}{2}(D + H\cos\omega_p t)(u_{n+1} - 2u_n + u_{n-1})$$
$$- \frac{1}{2}\eta\left[(u_{n+1} - u_n)^2(\dot{u}_{n+1} - \dot{u}_n) - (u_n - u_{n-1})^2(\dot{u}_n - \dot{u}_{n-1})\right] = 0, \quad (1.88)$$

where $u_n(t)$ describes the deviation of the nth resonator from its equilibrium, with $n = 1, \ldots, N$, and fixed boundary conditions $u_0 = u_{N+1} = 0$. Detailed arguments for the choice of terms introduced into the equations of motion are discussed in Ref. [60]. The terms include an elastic restoring force with both linear and cubic contributions (whose coefficients are both scaled to 1 as in our discussion of the single degree of freedom), a dc electrostatic nearest-neighbor coupling term with a small ac component responsible for the parametric excitation (with coefficients D and H respectively), and linear as well as cubic nonlinear dissipation terms. Both dissipation terms are assumed to depend on the difference of the displacements of nearest neighbors.

We consider here a slightly simpler and more general model for an array of coupled resonators to illustrate the approach. Motivated by the geometry of most experimental NEMS systems we suppose a line of identical resonators, although the generalization to two or three dimensions is straightforward. The simplest model is to take the equation of motion of each resonator to be as in Eq. (1.44) with the addition of a coupling term to its two neighbors. A simple choice would be to assume this coupling does not introduce additional dissipation, which we describe as *reactive*coupling. Elastic and electrostatic coupling might be predominantly of this type. After the usual scaling, the equations of motions would take the form

$$\ddot{u}_n + Q^{-1}\dot{u}_n + u_n^3 + (1 + H\cos\omega_P t)u_n + \eta u_n^2\dot{u}_n$$
$$+ \frac{1}{2}D(u_{n+1} - 2u_n + u_{n-1}) = 0 \quad (1.89)$$

where we do not take into account any direct drive for the purposes of the present section.

The equations of motion for particular experimental implementations might have different terms, although we expect all will have linear and nonlinear damping, linear coupling, and parametric drive. For example, to model the experimental setup of Buks and Roukes [37], LC supposed that both linear and nonlinear dissipation terms involved the difference of neighboring displacements, *i.e.*, the terms involving $\dot{u}_n$ in our equations of motion here (1.89) are replaced with terms involving $u_{n+1} - u_n$ in the equations of motion (1.88), used by LC. This was to describe the physics of electric current damping, with the currents driven by the varying capacitance between neighboring resonators depending on the change in separation and the fixed DC voltage. This effect seemed to be the dominant component of the dissipation in the Buks and Roukes experiments. Similarly the parametric drive $H \cos \omega_P t$ multiplied $(u_{n+1} - 2u_n + u_{n-1})$ in the equations of LC rather than u_n here, since the voltage between adjacent resonators was the quantity modulated, changing the electrostatic component of the spring constant. In more recent implementations [57], the electric current damping has been reduced, and the parametric drive is applied piezoelectrically directly to each resonator, so that the simpler equation (1.89) applies. The method of attack is the same in any case. We will illustrate the approach on the simpler equation, and refer the reader to LC for that more complicated model. An additional complication in a realistic model may be that the coupling is longer range than nearest neighbor. For example both electrostatic coupling and elastic coupling through the supports would have longer range components. The general method is the same for these additional effects, and the reader should be able to apply the approach to the model for their particular experimental implementation.

1.4.2
Calculating the Response of an Array

We calculate the response of the array to parametric excitation, again using secular perturbation theory. We suppose Q is large and take $\epsilon = Q^{-1}$ as the small expansion parameter. As in § 1.3 we take $H = \epsilon h$ but we also take $D = \epsilon d$ so that the width of the frequency band of eigenmodes is also small. This is not quite how LC treated the coupling, but we think the present approach is clearer, and it is equivalent up to the order of the expansion in ϵ that we require. We thank Eyal Kenig for pointing out this simplification.

The equations of motion are now

$$\ddot{u}_n + \epsilon \dot{u}_n + u_n^3 + (1 + \epsilon h \cos[(2 + \epsilon \Omega_P)t])u_n + \eta u_n^2 \dot{u}_n + \frac{1}{2}\epsilon d(u_{n+1} - 2u_n + u_{n-1}) = 0, \qquad n = 1, \ldots, N. \tag{1.90}$$

We expand $u_n(t)$ as a sum of standing wave modes with slowly varying amplitudes. The nature of the standing wave modes will depend on the conditions at the end of the line of oscillators. In the experiments of Buks and Roukes there where N mobile beams with a number of identical immobilized beams at each end. These conditions can be implemented in a nearest neighbor model by taking two additional resonators, u_0 and u_{N+1} and assuming

$$u_0 = u_{N+1} = 0. \tag{1.91}$$

The standing wave modes are then

$$u_n = \sin(nq_m) \qquad \text{with} \qquad q_m = \frac{m\pi}{N+1}, \quad m = 1, \ldots, N. \tag{1.92}$$

On the other hand for a line of N oscillators with free ends there is no force from outside the line. For the nearest neighbor model this can be imposed again by taking two additional oscillators, but now with the conditions

$$u_0 = u_1; \qquad u_N = u_{N+1}. \tag{1.93}$$

The standing wave modes are now

$$u_n = \cos\left[\left(n - \tfrac{1}{2}\right) q_m\right] \qquad \text{with} \qquad q_m = \frac{m\pi}{N}, \; m = 0, \ldots, N-1. \tag{1.94}$$

For our illustration we will take Eqs. (1.91) and (1.92). Thus we write

$$u_n(t) = \epsilon^{1/2} \frac{1}{2} \sum_{m=1}^{N} \left(A_m(T) \sin(nq_m) e^{it} + c.c. \right) + \epsilon^{3/2} u_n^{(1)}(t) + \ldots, \qquad n = 1, \ldots, N. \tag{1.95}$$

with q_m as in Eq. (1.92).

We substitute the trial solution (1.95) into the EOM term by term. Up to order $\epsilon^{3/2}$ we have:

$$\ddot{u}_n = \epsilon^{1/2} \frac{1}{2} \sum_m \sin(nq_m) \left([-A_m + 2i\epsilon A'_m] e^{it} + c.c. \right) + \epsilon^{3/2} \ddot{u}_n^{(1)}(t) + \ldots, \tag{1.96a}$$

$$\epsilon \dot{u}_n = \epsilon^{3/2} \frac{1}{2} \sum_m \sin(nq_m) \left(iA_m e^{it} + c.c. \right) + \ldots, \tag{1.96b}$$

$$\frac{1}{2} \epsilon d(u_{n+1} - 2u_n + u_{n-1}) = -\epsilon^{3/2} \frac{d}{2} \sum_m 2\sin^2\left(\frac{q_m}{2}\right) \sin(nq_m) \left(A_m e^{it} + c.c. \right) + \ldots \tag{1.96c}$$

$$u_n^3 = \epsilon^{3/2}\frac{1}{8}\sum_{j,k,l}\sin(nq_j)\sin(nq_k)\sin(nq_l)$$
$$\times\left(A_je^{it}+c.c.\right)\left(A_ke^{it}+c.c.\right)\left(A_le^{it}+c.c.\right)$$
$$=\epsilon^{3/2}\frac{1}{32}\sum_{j,k,l}\{\sin[n(-q_j+q_k+q_l)]+\sin[n(q_j-q_k+q_l)]$$
$$+\sin[n(q_j+q_k-q_l)]-\sin[n(q_j+q_k+q_l)]\}$$
$$\times\left\{A_jA_kA_le^{3it}+3A_jA_kA_l^*e^{it}+c.c.\right\}, \tag{1.96d}$$

and

$$\eta {u_n}^2\ddot{u}_n = \epsilon^{3/2}\frac{\eta}{32}\sum_{j,k,l}\{\sin[n(-q_j+q_k+q_l)]+\sin[n(q_j-q_k+q_l)]$$
$$+\sin[n(q_j+q_k-q_l)]-\sin[n(q_j+q_k+q_l)]\}$$
$$\times\left(A_je^{it}+c.c.\right)\left(A_ke^{it}+c.c.\right)\left(iA_le^{it}+c.c.\right). \tag{1.96e}$$

The order $\epsilon^{1/2}$ terms cancel, and at order $\epsilon^{3/2}$ we get N equations of the form

$$\ddot{u}_n^{(1)}+u_n^{(1)}=\sum_m\left(m^{th}\text{ secular term}\right)e^{it}+\text{other terms}, \tag{1.97}$$

where the left-hand sides are uncoupled linear harmonic oscillators, with a frequency unity. On the right-hand sides we have N secular terms which act to drive the oscillators $u_n^{(1)}$ at their resonance frequencies. As we did for all the single resonator examples, here too we require that all the secular terms vanish so that the $u_n^{(1)}$ remain finite, and thus obtain equations for the slowly varying amplitudes $A_m(T)$. To extract the equation for the mth amplitude $A_m(T)$ we make use of the orthogonality of the modes, multiplying all the terms by $\sin(nq_m)$ and summing over n. We find that the coefficient of the mth secular term, which is required to vanish, is given by

$$-2i\frac{dA_m}{dT}-iA_m+2d\sin^2\left(\frac{q_m}{2}\right)A_m-\frac{1}{2}hA_m^*e^{i\Omega_pT}$$
$$-\frac{3+i\eta}{16}\sum_{j,k,l}A_jA_kA_l^*\Delta_{jkl;m}^{(1)}=0, \tag{1.98}$$

where we have used the Δ function introduced by LC, defined in terms of Kronecker deltas as

$$\begin{aligned}\Delta^{(1)}_{jkl;m} &= \delta_{-j+k+l,m} - \delta_{-j+k+l,-m} - \delta_{-j+k+l,2(N+1)-m} \\ &+ \delta_{j-k+l,m} - \delta_{j-k+l,-m} - \delta_{j-k+l,2(N+1)-m} \\ &+ \delta_{j+k-l,m} - \delta_{j+k-l,-m} - \delta_{j+k-l,2(N+1)-m} \\ &- \delta_{j+k+l,m} + \delta_{j+k+l,2(N+1)-m} - \delta_{j+k+l,2(N+1)+m}\,,\end{aligned} \tag{1.99}$$

and have exploited the fact that it is invariant under any permutation of the indices j, k, and l.

The function $\Delta^{(2)}_{jkl;m}$, also defined by LC, is not needed for our simplified model. The Δ function ensures the conservation of lattice momentum – the conservation of momentum to within the non-uniqueness of the specification of the normal modes due to the fact that $\sin(nq_m) = \sin(nq_{2k(N+1)\pm m})$ for any integer k. The first Kronecker delta in each line is a condition of direct momentum conservation, and the other two are the so-called umklapp conditions where only lattice momentum is conserved.

As for the single resonator, we again try a steady-state solution, this time of the form

$$A_m(T) = a_m e^{i\left(\frac{\Omega_P}{2}\right)T}, \tag{1.100}$$

so that the solutions to the EOM, after substitution of Eq. (1.100) into Eq. (1.95), become

$$u_n(t) = \epsilon^{1/2}\frac{1}{2}\sum_m \left(a_m \sin(nq_m) e^{i\left(1+\frac{\epsilon\Omega_P}{2}\right)t} + c.c.\right) + O(\epsilon^{3/2}), \tag{1.101}$$

where all modes are oscillating at half the parametric excitation frequency.

Substituting the steady state solution (1.100) into the Eqs. (1.98) for the time-varying amplitudes $A_m(T)$, we obtain the equations for the time-independent complex amplitudes a_m

$$\left[\Omega_P + 2d\sin^2\left(\frac{q_m}{2}\right) - i\right] a_m - \frac{h}{2}a_m^* - \frac{3+i\eta}{16}\sum_{j,k,l} a_j a_k a_l^* \Delta^{(1)}_{jkl;m} = 0\,. \tag{1.102}$$

Note that the first two terms on the left-hand side indicate that the linear resonance frequency is not obtained for $\Omega_P = 0$, but rather for $\Omega_P + 2d\sin^2(q_m/2) = 0$. In terms of the unscaled parameters, this implies that the resonance frequency of the mth mode is $\omega_m = 1 - D\sin^2(q_m/2)$, which to within a correction of $O(\epsilon^2)$ is the same as the expected dispersion relation

$$\omega_m^2 = 1 - 2D\sin^2\left(\frac{q_m}{2}\right). \tag{1.103}$$

Equation (1.102) is the main result of the calculation. We have managed to replace N coupled differential equations (1.89) for the resonator coordinates $u_n(t)$ by N coupled algebraic equations (1.102) for the time-independent mode amplitudes a_m. All that remains, in order to obtain the overall collective response of the array as a function of the parameters of the original EOM, is to solve these coupled algebraic equations.

First, one can easily verify that for a single oscillator $(N=j=k=l=m=1)$, the general equation (1.102) reduces to the single-oscillator equation (1.73), we derived in § 1.3.4, as $\Delta_{111;1} = 4$. Next, one can also see that the trivial solution, $a_m = 0$ for all m, always satisfies the equations, though, as we have seen in the case of a single oscillator, it is not always a stable solution. Finally, one can also verify that whenever for a given m, $\Delta^{(1)}_{mmm;j} = 0$ for all $j \neq m$, then a single-mode solution exists with $a_m \neq 0$ and $a_j = 0$ for all $j \neq m$. These single-mode solutions have the same type of elliptical shape of the single-oscillator solution given in Eq. (1.74). Note that generically $\Delta^{(1)}_{mmm;m} = 3$, except when umklapp conditions are satisfied.

Additional solutions, involving more than a single mode, exist in general but are hard to obtain analytically. LC calculated these multi-mode solutions for the case of two and three oscillators for the model they considered by finding the roots of the coupled algebraic equations numerically. We show some of their results to illustrate the type of behavior that occurs, although the precise details will be slightly different.

1.4.3 The Response of Very Small Arrays – Comparison of Analytics and Numerics

In Fig. 1.13 we show the solutions for the response intensity of two resonators as a function of frequency, for a particular choice of the equation parameters. The top graph shows the square of the amplitude of the antisymmetric mode a_2, whereas the bottom graph shows the square of the amplitude of the symmetric mode a_1. Solid curves indicate stable solutions and dashed curves indicate unstable solutions. Two elliptical single-mode solution branches, similar to the response of a single resonator shown in Fig. 1.10 are easily spotted. These branches are labeled by S_1 and S_2. LC give the analytical expressions for these two solution branches. In addition, there are two double-mode solution branches, labeled D_1 and D_2, involving the excitation of both modes simultaneously. Note that the two branches of double-mode solutions intersect at a point where they switch their stability.

With two resonators there are regions in frequency where three stable solutions can exist. If all the stable solution branches are accessible experimentally then the observed effects of hysteresis might be more complex than in the sim-

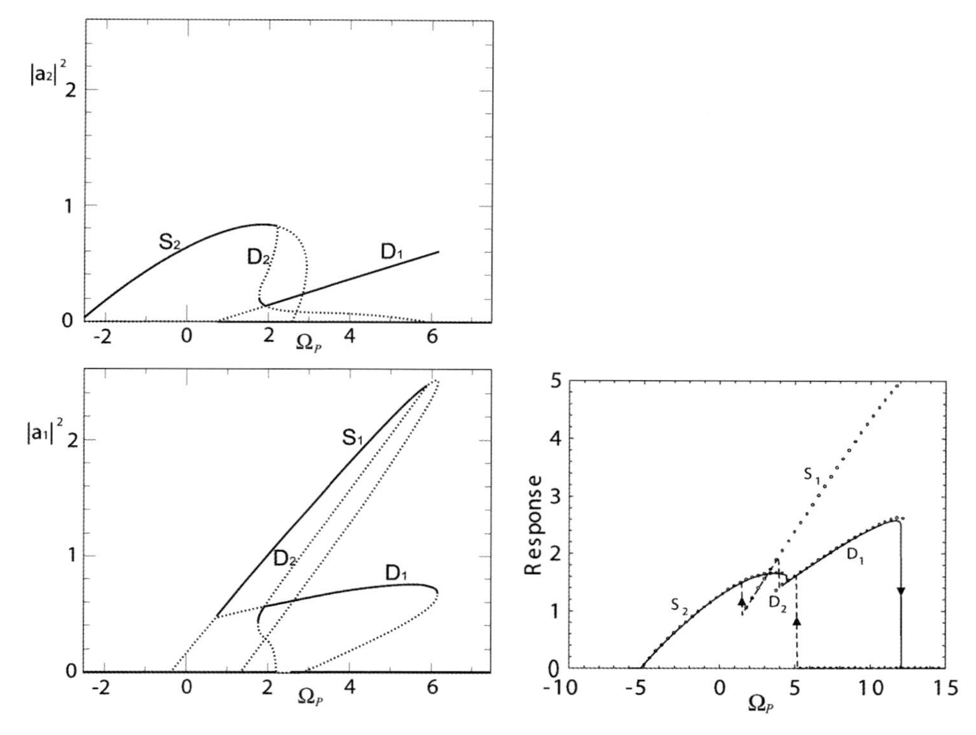

Fig. 1.13 Two resonators. Left: Response intensity as a function of frequency Ω_P, for a particular choice of the equation parameters. The top graph shows $|a_2|^2$, the bottom graph shows $|a_1|^2$. Solid curves indicate stable solutions and dashed curves indicate unstable solutions. The two elliptical single-mode solution branches are labeled S_1 and S_2. The two double-mode solution branches are labeled D_1 and D_2. Right: Comparison of stable solutions, obtained analytically (small circles), with a numerical integration of the equations of motion (solid curve: frequency swept up; dashed curve: frequency swept down) showing hysteresis in the response. Plotted is the averaged response intensity, defined in Eq. (1.104). Branch labels correspond to those on the left.

ple case of a single resonator. This is demonstrated on the right-hand side of Fig. 1.13 where the analytical solutions are compared with a numerical integration of the differential equations of motion (1.89) for two resonators. The response intensity, plotted here, is the time and space averages of the square of the resonator displacements

$$I = \frac{1}{N}\sum_{n=1}^{N}\left\langle u_n^2 \right\rangle, \tag{1.104}$$

where the angular brackets denote time average, and here $N = 2$. A solid curve shows the response intensity for frequency swept upwards, and a dashed curve shows the response intensity for frequency swept downwards. Small circles show the analytical response intensity, for the stable regions of the four solution branches shown in Fig. 1.13. With the analytical solution in the background, one can easily understand all the discontinuous jumps, as well as the hysteresis effects, that are obtained in the numerical solution of the equations of motion. Note that the S_1 branch is missed in the upwards frequency sweep and is only accessed by the system in the downwards frequency sweep. One could trace the whole stable region of the S_1 branch by changing the sweep direction after jumping onto the branch, thereby climbing all the way up to the end of the S_1 branch and then falling onto the tip of the D_1 branch or to zero. These kinds of changes in the direction of the sweep whenever one jumps onto a new branch are essential if one wants to trace out as much of the solution as possible – whether in real experiments or in numerical simulations.

1.4.4
Response of Large Arrays – Numerical Simulation

LC integrated the equations of motion (1.88) numerically for an array of $N = 67$ resonators. The results for the response intensity (1.104) as a function of the unscaled parametric drive frequency ω_p are shown in Fig. 1.14. These results must be considered illustrative only, because the structure of the response branches will vary with changes to the model, and will also depend

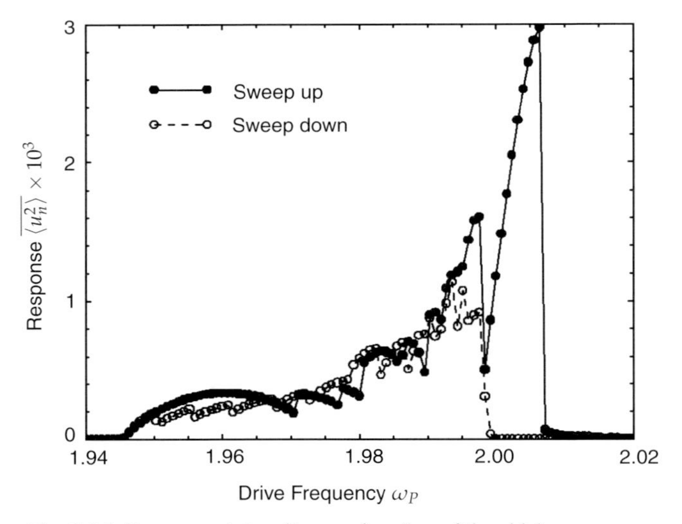

Fig. 1.14 Response intensity as a function of the driving frequency ω_p for $N = 67$ parametrically-driven resonators (solid curve: frequency swept up; dashed curve: frequency swept down). The response intensity is defined in Eq. (1.104). The response curve was obtained through numerical integration of the equations of motion (1.88).

strongly on the chosen equation parameters. First of all, as in the case of a small number of beams, the overall height and width of individual response branches depend on the strength of the drive h, and on the nonlinear dissipation coefficient η. Furthermore, for example, if the coupling strength D is increased so that the width of the frequency response band become much larger than N times the width of a single-mode response, then very few multi-mode solutions exist, if any.

A number of the important features should be pointed out in the response. We concentrate on the solid curve in the figure, which is for frequency swept upwards. First, the response intensity shows features that span a range of frequencies that is large compared with the mode spacing (which is about 0.0006 for the parameters used). The reason for this is that as we follow a particular solution we skip over many others, as we saw for the S1 branch in the two-resonator case. Second, the variation of the response with frequency shows abrupt jumps, particularly on the high frequency side of the features as the frequency is raised. This happens as we reach saddle-node or other types of bifurcations where we lose the stability of the solution branch or the branch ends altogether. Third, the response extends to frequencies higher than the band edge for the linear modes, which would give a response only up to $\omega_p = 2.0$. This happens simply due to the positive Duffing parameter which causes frequency pulling to the right. Note that the downwards sweep is able to access additional stable solution branches that were missed in the upwards sweep. There is also no response above $\omega_p = 2.0$ in this case. This is because the zero displacement state is stable for $\omega_p > 2.0$, and the system will remain in this state as the frequency is lowered, unless a large enough disturbance kicks it onto another of the solution branches. The hysteresis on reversing the frequency sweep was not looked at in any experiment, and it would be interesting to test this prediction of LC in the future.

1.5 Amplitude Equation Description for Large Arrays

We finish this review by describing the approach used by BCL [61,63] to obtain analytical results for large arrays by approaching them from the continuous limit of infinitely-many degrees of freedom. We only summarize the main results of BCL and encourage the reader, who by now has all the required background, to refer to BCL [61] and to Kenig et al. [62] for details of the derivation and for thorough discussions of the results and their experimental consequences. We note that BCL studied the original system of equations (1.88), where both the parametric excitation and the damping are introduced in terms of the difference variables $u_{n+1} - u_n$. We stick to this model here, and leave it to the reader as an exercise to generalize the BCL derivation for the more general model equations (1.89) that we used in the previous section.

A novel feature of the parametrically-driven instability is that the bifurcation to standing waves switches from supercritical (continuous) to subcritical (discontinuous) at a wave number at or close to the critical one, for which the required parametric driving force is minimum. This changes the form of the amplitude equation that describes the onset of the parametrically-driven waves so that it no longer has the standard "Ginzburg–Landau" form [46]. The central result of BCL is this new scaled amplitude equation (1.112), governed by a single control parameter, that captures the slow dynamics of the coupled resonators just above the onset of parametric oscillations, including this unusual bifurcation behavior. BCL confirmed the behavior numerically and made suggestions for testing it experimentally. Kenig et al. [62] have extended the investigation of the amplitude equation to include such situations as time-dependent ramps of the drive amplitude, as opposed to the standard quasistatic sweeps of the control parameters. Although our focus here is on parametrically-driven NEMS & MEMS resonators, we should emphasize that the amplitude equation of BCL which we describe here should also apply to other parametrically-driven wave systems with weak nonlinear damping.

1.5.1 Amplitude Equations for Counter Propagating Waves

BCL scaled the equations of motion (1.88), as did Lifshitz and Cross [60], without assuming *a priori* that the coupling D is small, thus the scaled equations of motion that they solved were

$$\begin{aligned}\ddot{u}_n + u_n + u_n^3 - \frac{1}{2}\epsilon(\dot{u}_{n+1} - 2\dot{u}_n + \dot{u}_{n-1}) \\ + \frac{1}{2}\left[D + \epsilon h \cos(2\omega_p t)\right](u_{n+1} - 2u_n + u_{n-1}) \\ - \frac{1}{2}\eta\left[(u_{n+1} - u_n)^2(\dot{u}_{n+1} - \dot{u}_n) - (u_n - u_{n-1})^2(\dot{u}_n - \dot{u}_{n-1})\right] = 0,\end{aligned} \tag{1.105}$$

Note also the way in which the pump frequency is specified as $2\omega_p$ in the argument of the cosine term, with an explicit factor of two (unlike what we did in Section 1.4), and also without making any assumptions at this point regarding its deviation from twice the resonance. We also remind the reader that this, and all other frequencies, are measured in terms of the natural frequency of a single resonator which has been scaled to 1. The first step in treating this system of equations analytically, is to introduce a continuous displacement field $u(x,t)$, and slow spatial and temporal scales, $X = \epsilon x$ and

$T = \epsilon t$. One then tries a solution in terms of a pair of counter-propagating plane waves, at half the pump frequency, which is a natural first guess in continuous parametrically-driven systems, such as Faraday waves [46],

$$u(x,t) = \epsilon^{1/2}\left[\left(A_+(X,T)e^{-iq_p x} + A_-^*(X,T)e^{iq_p x}\right)e^{i\omega_p t} + c.c.\right] + \epsilon^{3/2}u^{(1)}(x,t,X,T) + \ldots, \tag{1.106}$$

where q_p and ω_p are related through the dispersion relation (1.103)

$$\omega_p^2 = 1 - 2D\sin^2\left(\frac{q_p}{2}\right). \tag{1.107}$$

By substituting this ansatz (1.106) into the equations of motion (1.105) and applying a solvability condition on the terms of order $\epsilon^{3/2}$, BCL obtained a pair of coupled amplitude equations for the counter-propagating wave amplitudes $A_\pm$

$$\frac{\partial A_\pm}{\partial T} \pm v_g \frac{\partial A_\pm}{\partial X} = -\sin^2\left(\frac{q_p}{2}\right)A_\pm \mp i\frac{h}{2\omega_p}\sin^2\left(\frac{q_p}{2}\right)A_\mp - \left(4\eta\sin^4\left(\frac{q_p}{2}\right) \mp i\frac{3}{2\omega_p}\right)\left(|A_\pm|^2 + 2|A_\mp|^2\right)A_\pm, \tag{1.108}$$

where the upper signs (lower signs) give the equation for A_+ (A_-) and

$$v_g = \frac{\partial \omega_p}{\partial q_p} = -\frac{D\sin(q_p)}{2\omega_p} \tag{1.109}$$

is the group velocity. This equation is the extension of Eq. (1.47) to many coupled resonators, only that now the parametric drive couples amplitudes of the two counter propagating waves A_+ and A_- instead of coupling A and A^*. A detailed derivation of the amplitude equations (1.108) can be found in Refs. [61,63]. We should note that similar equations were previously derived for describing Faraday waves [64,65].

By linearizing the amplitude equations (1.108) about the zero solution ($A_+ = A_- = 0$) we find that the linear combination of the two amplitudes that first becomes unstable at $h = h_c \equiv 2\omega_p$ is $B \propto (A_+ - iA_-)$ – representing the emergence of a standing wave with a temporal phase of $\pi/4$ relative to the drive – while the orthogonal linear combination of the amplitudes decays exponentially and does not participate in the dynamics at onset. Thus, just above threshold a single amplitude equation should suffice, describing this standing wave pattern. We describe the derivation of this equation in the next section.

1.5.2
Reduction to a Single Amplitude Equation

Because, as we saw in Section 1.3.4, nonlinear dissipation plays an important role in the saturation of the response to parametric excitation, it is natural to try to keep a balance between the strength of this nonlinearity and the amount by which we drive the system above threshold. Assuming that the nonlinear damping is weak, we use it to define a second small parameter $\delta = \sqrt{\eta}$. This particular definition turns out to be useful if we then scale the reduced driving amplitude $(h - h_c)/h_c$ linearly with δ, defining a scaled reduced driving amplitude r by letting $(h - h_c)/h_c \equiv r\delta$. We can then treat the initial linear combination of the two amplitudes in Eq. (1.108) that becomes unstable by introducing a second ansatz,

$$\begin{pmatrix} A_+ \\ A_- \end{pmatrix} = \delta^{1/4} \begin{pmatrix} 1 \\ i \end{pmatrix} B(\xi, \tau) + \delta^{3/4} \begin{pmatrix} w^{(1)}(X, T, \xi, \tau) \\ v^{(1)}(X, T, \xi, \tau) \end{pmatrix} + \delta^{5/4} \begin{pmatrix} w^{(2)}(X, T, \xi, \tau) \\ v^{(2)}(X, T, \xi, \tau) \end{pmatrix}, \tag{1.110}$$

where $\xi = \delta^{1/2} X$ and $\tau = \delta T$. Substitution of this ansatz allows one to obtain the correction of the solution at order $\delta^{3/4}$

$$\begin{pmatrix} w^{(1)} \\ v^{(1)} \end{pmatrix} = \frac{1}{2\sin^2(q_p/2)} \left(-v_g \frac{\partial B}{\partial \xi} + i \frac{9}{2\omega_p} |B|^2 B \right) \begin{pmatrix} 1 \\ -i \end{pmatrix}, \tag{1.111}$$

after which a solvability condition applied to the terms of order $\delta^{5/4}$ yields an equation for the field $B(\xi, \tau)$, which after scaling takes the form

$$\frac{\partial B}{\partial \tau} = rB + \frac{\partial^2 B}{\partial \xi^2} + i\frac{2}{3} \left(4|B|^2 \frac{\partial B}{\partial \xi} + B^2 \frac{\partial B^*}{\partial \xi} \right) - 2|B|^2 B - |B|^4 B. \tag{1.112}$$

This is the BCL amplitude equation. It is governed by a single control parameter, the reduced drive amplitude r, and captures the slow dynamics of the coupled resonators just above the onset of parametric oscillations. The reader is encouraged to consult Ref. [61] for a more detailed account of the derivation of the BCL equation. The form of Eq. (1.112) is also applicable to the onset of parametrically driven standing waves in continuum systems with weak nonlinear damping, and combines in a single equation a number of effects studied previously [64–69].

1.5.3
Single Mode Oscillations

Now that this novel amplitude equation has been derived by BCL it can be used to study a variety of dynamical solutions, ranging from simple single-mode to more complicated nonlinear extended solutions, and after slight modifications also the dynamics of localized solutions. BCL used the amplitude equation to study the stability of single-mode steady-state solutions,

$$B = b_k e^{-ik\xi}, \tag{1.113}$$

i.e. standing-wave solutions that consist of a single sine-wave pattern with one of the allowed wave vectors q_m. The wave vector k gives, in some scaled units, the difference between the wave vector q_p, determined by the pump frequency through the dispersion relation, and the wave vector $q_m = m\pi/(N+1)$, $m = 1,\ldots,N$, of the actual mode that is selected by the system.

A number of interesting results easily pop out if we simply substitute the single-mode solution (1.113) into the BCL amplitude equation (1.112). From the linear terms in the amplitude equation we find, as expected, that for $r > k^2$ the zero-displacement solution is unstable to small perturbations of the form of Eq. (1.113), defining the parabolic neutral stability curve, shown as a dashed

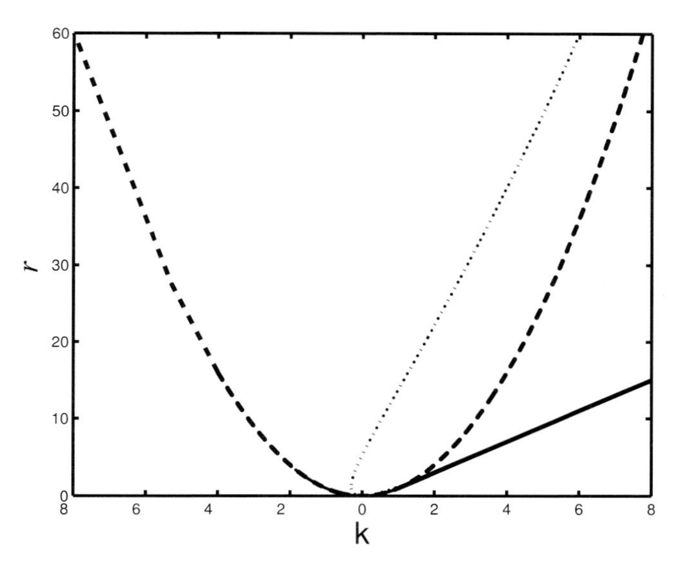

Fig. 1.15 Stability boundaries of the single-mode solution of Eq. (1.112) in the r *vs.* k plane. Dashed line: neutral stability boundary below which the zero-state is stable. Dotted line: stability boundary of the single-mode solution (1.113) above which the array experiences an Eckhaus instability and switches to one of the other single mode solutions. For $k > 1$ the bifurcation from zero displacement becomes subcritical and the lower stability boundary is the locus of saddle-node bifurcations (solid line).

line in Fig. 1.15. The nonlinear gradients and the cubic term take the simple form $2(k-1)|b_k|^2 b_k$. For $k < 1$ these terms immediately act to saturate the growth of the amplitude assisted by the quintic term. Standing waves therefore bifurcate *supercritically* from the zero-displacement state. For $k > 1$ the cubic terms act to increase the growth of the amplitude, and saturation is achieved only by the quintic term. Standing waves therefore bifurcate *subcritically* from the zero-displacement state. The saturated amplitude $|b_k|$, obtained by setting Eq. (1.112) to zero, is given by

$$|b_k|^2 = (k-1) \pm \sqrt{(k-1)^2 + (r-k^2)} \geq 0. \tag{1.114}$$

In Fig. (1.16) we plot $|b_k|^2$ as a function of the reduced driving amplitude r for three different wave number shifts k. The solid (dashed) lines are the stable (unstable) solutions of Eq. (1.114). The circles were obtained by numerical integration of the equations of motion (1.105). For each driving amplitude, the Fourier components of the steady state solution were computed to verify that only single modes are found, suggesting that in this regime of parameters only these states are stable.

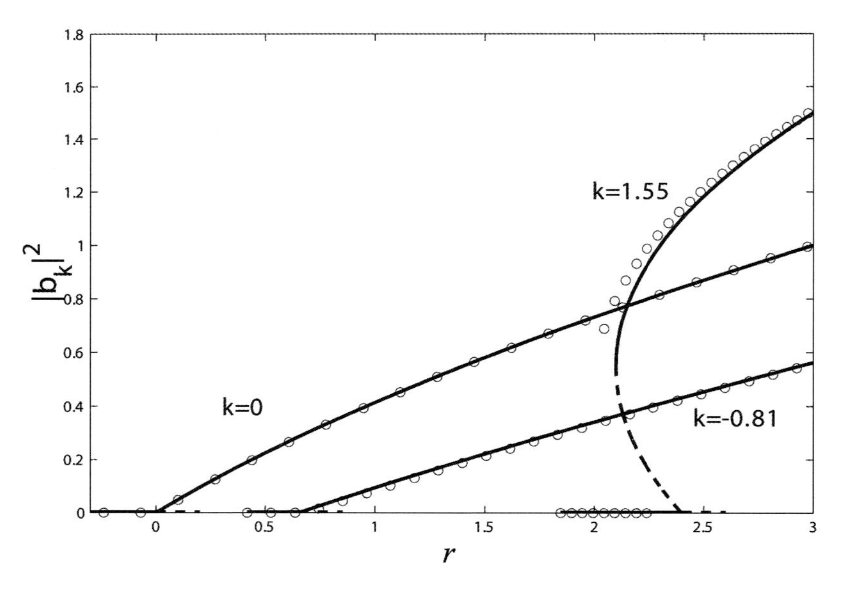

Fig. 1.16 Response of the resonator array plotted as a function of reduced amplitude r for three different scaled wave number shifts: $k = 0$ and $k = -0.81$, which bifurcate supercritically, and $k = 1.55$ which bifurcates subcritically showing clear hysteresis. Solid and dashed lines are the positive and negative square root branches of the calculated response in (1.114), the latter clearly unstable. Open circles are numerical values obtained by integration of the equations of motion (1.105), with $D = 0.25$, $\omega_p = 0.767445$, $\epsilon = 0.01$, and $\eta = 0.1$.

BCL showed the power of the amplitude equation in predicting the first single-mode solution that should appear at onset as well as the sequence of Eckhaus instabilities that switch to other single-mode solutions as the reduced drive amplitude r is quasistatically increased. Kenig et al. used the amplitude equation to analyze more generally the question of pattern selection – predicting which oscillating pattern will be observed under particular experimental conditions from among all the stable steady-state solutions that the array of resonators can choose from. In particular they have considered experimental situations in which the drive amplitude r is changed abruptly or swept at rates that are faster than typical transient times. In all cases the predictions of the amplitude equation are confirmed with numerical simulations of the original equations of motion (1.105). Experimental confirmation of these predictions is still not available.

Acknowledgments

We wish to thank the students at Tel Aviv University, Yaron Bromberg and Eyal Kenig, who have worked with us on developing and then using the amplitude equation for the treatment of large arrays of parametrically-driven Duffing resonators. We wish to thank our experimental colleagues, Eyal Buks, Rassul Karabalin, Inna Kozinsky, and Michael Roukes, for many fruitful interactions. We also wish to thank Andrew Cleland, Harry Dankowicz, Oleg Kogan, Steve Shaw, and Kimberly Turner for stimulating discussions. This work is funded by the U.S.-Israel Binational Science Foundation (BSF) through Grant No. 2004339, by the U.S. National Science Foundation (NSF) through Grant No. DMR-0314069, and by the Israeli Ministry of Science.

References

1 M. L. Roukes. Plenty of room indeed. *Scientific American*, 285:42–49, 2001.

2 M. L. Roukes. Nanoelectromechanical systems face the future. *Physics World*, 14:25–31, 2001.

3 H. G. Craighead. Nanoelectromechanical systems. *Science*, 290:1532–1535, 2000.

4 Andrew N. Cleland. *Foundations of Nanomechanics*. Springer, Berlin, 2003.

5 K. L. Ekinci and M. L. Roukes. Nanoelectromechanical systems. *Review of Scientific Instruments*, 76:061101, 2005.

6 K. L. Ekinci, X. M. H. Huang, and M. L. Roukes. Ultrasensitive nanoelectromechanical mass detection. *Appl. Phys. Lett.*, 84:4469–4471, 2004.

7 B. Ilic, H. G. Craighead, S. Krylov, W. Senaratne, C. Ober, and P. Neuzil. Attogram detection using nanoelectromechanical oscillators. *Journal of Applied Physics*, 95:3694–3703, 2004.

8 Y.T. Yang, C. Callegari, X.L. Feng, K.L. Ekinci, and M.L. Roukes. Zeptogram-scale nanomechanical mass sensing. *Nano Letters*, 6:583–586, 2006.

9 D Rugar, R Budakian, H J Mamin, Chui, and B W. Single spin detection by magnetic

resonance force microscopy. *Nature*, 430:329–332, 2004.

10 Andrew N. Cleland and Michael L. Roukes. A nanometer-scale mechanical electrometer. *Nature*, 392:160, 1998.

11 Andrew N. Cleland and Michael L. Roukes. Nanoscale mechanics. In *Proceedings of the 24th International Conference on the Physics of Semiconductors*. World Scientific, 1999.

12 Keith C. Schwab, E. A. Henriksen, J. M. Worlock, and Michael L. Roukes. Measurement of the quantum of thermal conductance. *Nature*, 404:974–977, 2000.

13 Miles P. Blencowe. Quantum electromechanical systems. *Physics Reports*, 395:159–222, 2004.

14 Keith C. Schwab and Michael L. Roukes. Putting mechanics into quantum mechanics. *Physics Today*, 58(7):36–42, 2005.

15 Kimberly L. Turner, Scott A. Miller, Peter G. Hartwell, Noel C. MacDonald, Steven H. Strogatz, and Scott G. Adams. Five parametric resonances in a microelectromechanical system. *Nature*, 396:149–152, 1998.

16 Eyal Buks and Michael L. Roukes. Metastability and the casimir effect in micromechanical systems. *Europhys. Lett.*, 54:220, 2001.

17 Dominik V. Scheible, Artur Erbe, Robert H. Blick, and Gilberto Corso. Evidence of a nanomechanical resonator being driven into chaotic response via the ruelle–takens route. *Applied Physics Letters*, 81:1884–1886, 2002.

18 Wenhua Zhang, Rajashree Baskaran, and Kimberly L. Turner. Effect of cubic nonlinearity on auto-parametrically amplified resonant mems mass sensor. *Sensors and Actuators A*, 102:139–150, 2002.

19 Wenhua Zhang, Rajashree Baskaran, and Kimberly Turner. Tuning the dynamic behavior of parametric resonance in a micromechanical oscillator. *Applied Physics Letters*, 82:130–132, 2003.

20 Min-Feng Yu, Gregory J. Wagner, Rodney S. Ruoff, and Mark J. Dyer. Realization of parametric resonances in a nanowire mechanical system with nanomanipulation inside a scanning electron microscope. *Physical Review B*, 66:073406, 2002.

21 J. S. Aldridge and A. N. Cleland. Noise-enabled precision measurements of a duffing nanomechanical resonator. *Physical Review Letters*, 94:156403, 2005.

22 A. Erbe, H. Krommer, A. Kraus, R. H. Blick, G. Corso, and K. Richter. Mechanical mixing in nonlinear nanomechanical resonators. *Applied Physics Letters*, 77:3102–3104, 2000.

23 D. Rugar and P. Grütter. Mechanical parametric amplification and thermomechanical noise squeezing. *Phys. Rev. Lett.*, 67:699, 1991.

24 Dustin W. Carr, Stephane Evoy, Lidija Sekaric, H. G. Craighead, and J. M. Parpia. Parametric amplification in a torsional microresonator. *Applied Physics Letters*, 77:1545–1547, 2000.

25 Dustin W. Carr, S. Evoy, L. Sekaric, H. G. Craighead, and J. M. Parpia. Measurement of mechanical resonance and losses in nanometer scale silicon wires. *Applied Physics Letters*, 75:920–922, 1999.

26 Eyal Buks and Bernard Yurke. Mass detection with a nonlinear nanomechanical resonator. *Physical Review E*, 74:046619, 2006.

27 R. Almog, S. Zaitsev, O. Shtempluck, and E. Buks. High intermodulation gain in a micromechanical duffing resonator. *Appl. Phys. Lett.*, 88:213509, 2006.

28 X. L. Feng, Rongrui He, Peidong Yang, , and M. L. Roukes. Very high frequency silicon nanowire electromechanical resonators. *Nano Letters*, 7:1953–1959, 2007.

29 Robert B. Reichenbach, Maxim Zalalutdinov, Keith L. Aubin, Richard Rand, Brian H. Houston, Jeevak M. Parpia, and Harold G. Craighead. Third-order intermodulation in a micromechanical thermal mixer. *J. MEMS*, 14:1244–1252, 2005.

30 A. Husain, J. Hone, Henk W. Ch. Postma, X. M. H. Huang, T. Drake, M. Barbic, A. Scherer, , and M. L. Roukes. Nanowire-based very-high-frequency electromechanical resonator. *Appl. Phys. Lett.*, 83:1240–1242, 2003.

31 Barry E. DeMartini, Jeffrey F. Rhoads, Kimberly L. Turner, Steven W. Shaw, and Jeff Moehlis. Linear and nonlinear tuning of parametrically excited mems oscillators. *J. MEMS*, 16:310–318, 2007.

32 Vera Sazonova, Yuval Yaish, Hande Üstünel, David Roundy, Tomás A. Arias, and Paul L. McEuen. A tunable carbon nanotube electromechanical oscillator. *Nature*, 431:284–287, 2004.

33 X. M. H. Huang, C. A. Zorman, M. Mehregany, and Michael L. Roukes. *Nature*, 421:496, 2003.

34 A. N. Cleland and M. R. Geller. Superconducting qubit storage and entanglement with nanomechanical resonators. *Physical Review Letters*, 93:070501, 2004.

35 H. B. Peng, C. W. Chang, S. Aloni, T. D. Yuzvinsky, and A. Zettl. Ultrahigh frequency nanotube resonators. *Physical Review Letters*, 97:087203, 2006.

36 M. C. Cross, A. Zumdieck, Ron Lifshitz, and J. L. Rogers. Synchronization by nonlinear frequency pulling. *Physical Review Letters*, 93:224101, 2004.

37 E. Buks and M. L. Roukes. Electrically tunable collective response in a coupled micromechanical array. *J. MEMS*, 11:802–807, 2002.

38 M. Sato, B. E. Hubbard, A. J. Sievers, B. Ilic, D. A. Czaplewski, and H. G. Craighead. Observation of locked intrinsic localized vibrational modes in a micromechanical oscillator array. *Physical Review Letters*, 90:044102, 2003.

39 M. Sato, B. E. Hubbard, A. J. Sievers, B. Ilic, D. A. Czaplewski, and Harold G. Craighead. Studies of intrinsic localized vibrational modes in micromechanical oscillator arrays. *Chaos*, 13:702–715, 2003.

40 M. Sato, B. E. Hubbard, A. J. Sievers, B. Ilic, and Harold G. Craighead. Optical manipulation of intrinsic localized vibrational energy in cantilever arrays. *Europhys. Lett.*, 66:318–323, 2004.

41 L. D. Landau and E. M. Lifshitz. *Theory of Elasticity*. Butterworth-Heinemann, Oxford, 3rd edition, 1986. §20 & 25.

42 B. Yurke, D. S. Greywall, A. N. Pargellis, and P. A. Busch. Theory of amplifier-noise evasion in an oscillator employing a nonlinear resonator. *Phys. Rev. A*, 51:4211–4229, 1995.

43 H. W. Ch. Postma, I. Kozinsky, A. Husain, and M. L. Roukes. Dynamic range of nanotube- and nanowire-based electromechanical systems. *Appl. Phys. Lett.*, 86:223105, 2005.

44 L. N. Hand and J. D. Finch. *Analytical Mechanics*, chapter 10. Cambridge Univ. Press, Cambridge, 1998.

45 Steven H. Strogatz. *Nonlinear dynamics and chaos*, chapter 7. Addison-Wesley, Reading MA, 1994.

46 M. C. Cross and P. C. Hohenberg. Pattern formation outside of equillibrium. *Rev. Mod. Phys.*, 65:851–1112, 1993.

47 Evoy, Carr, Sekaric, Olkhovets, Parpia, and Craighead. *J. Appl. Phys.*, 86:6072, 1999.

48 Stav Zaitsev, Ronen Almog, Oleg Shtempluck, and Eyal Buks. Nonlinear dynamics in nanomechanical oscillators. *Proceedings of the 2005 International Conference on MEMS,NANO and Smart Systems*, pages 387–391, 2005.

49 R. Almog, S. Zaitsev, O. Shtempluck, and E. Buks. Noise squeezing in a nanomechanical duffing resonator. *Physical Review Letters*, 98:078103, 2007.

50 Inna Kozinsky, Henk W. Ch. Postma, Oleg Kogan, A. Husain, and Michael L. Roukes. Basins of attraction of a nonlinear nanomechanical resonator. *Phys. Rev. Lett.*, 99:207201, 2007.

51 Itamar Katz, Alex Retzker, Raphael Straub, and Ron Lifshitz. Signatures for a classical to quantum transition of a driven nonlinear nanomechanical resonator. *Phys. Rev. Lett.*, 99:040404, 2007.

52 I. Kozinsky, H. W. Ch. Postma, I. Bargatin, and M. L. Roukes. Tuning nonlinearity, dynamic range, and frequency of nanomechanical resonators. *Appl. Phys. Lett.*, 88:253101, 2006.

53 L. D. Landau and E. M. Lifshitz. *Mechanics*. Butterworth-Heinemann, Oxford, 3rd edition, 1976. §27.

54 Rajashree Baskaran and Kimberly L Turner. Mechanical domain coupled mode parametric resonance and amplification in a torsional mode micro electro mechanical oscillator. *J. Micromech. Microeng.*, 13:701–707, 2003.

55 Jeffrey F. Rhoads, Steven W. Shaw, Kimberly L. Turner, Jeff Moehlis Barry E. DeMartini, and Wenhua Zhang.

Generalized parametric resonance in electrostatically actuated microelectromechanical oscillators. *J. Sound Vib.*, 296:797–829, 2006.

56 Jeffrey F Rhoads, Steven W Shaw, and Kimberly L Turner. The nonlinear response of resonant microbeam systems with purely-parametric electrostatic actuation. *J. Micromech. Microeng.*, 16:890–899, 2006.

57 S. C. Masmanidis, R. B. Karabalin, I. De Vlaminck, G. Borghs, M. R. Freeman, and M. L. Roukes. Multifunctional nanomechanical systems via tunably-coupled piezoelectric actuation. *Science*, 317:780–783, 2007.

58 E. Buks and M. L. Roukes. Electrically tunable collective response in a coupled micromechanical array. *J. MEMS*, 11:802–807, 2002.

59 Jeff Moehlis. Private communication.

60 Ron Lifshitz and M. C. Cross. Response of parametrically driven nonlinear coupled oscillators with application to micromechanical and nanomechanical resonator arrays. *Physical Review B*, 67:134302, 2003.

61 Yaron Bromberg, M. C. Cross, and Ron Lifshitz. Response of discrete nonlinear systems with many degrees of freedom. *Physical Review E*, 73:016214, 2006.

62 Eyal Kenig, Ron Lifshitz, and Michael C. Cross. Pattern selection in parametrically-driven arrays of nonlinear micromechanical or nanomechanical resonators. *Preprint*, 2007.

63 Yaron Bromberg. Response of nonlinear systems with many degrees of freedom. Master's thesis, Tel Aviv University, 2004.

64 A. B. Ezerskiĭ, M. I. Rabinovich, V. P. Reutov, and I. M. Starobinets. Spatiotemporal chaos in the parametric excitation of capillary ripple. *Zh. Eksp. Teor. Fiz.*, 91:2070–2083, 1986. [Sov. Phys. JETP **64**, 1228 (1986)].

65 S. T. Milner. Square patterns and secondary instabilities in driven capillary waves. *J. Fluid Mech.*, 225:81–100, 1991.

66 H. Riecke. Stable wave-number kinks in parametrically excited standing waves. *Europhys. Lett.*, 11:213–218, 1990.

67 Robert J. Deissler and Helmut R. Brand. Effect of nonlinear gradient terms on breathing localized solutions in the quintic complex Ginzburg–Landau equation. *Phys. Rev. Lett.*, 81:3856–3859, 1998.

68 Peilong Chen and Kuo-An Wu. Subcritical bifurcations and nonlinear ballons in faraday waves. *Phys. Rev. Lett.*, 85:3813–3816, 2000.

69 Peilong Chen. Nonlinear wave dynamics in faraday instabilities. *Physical Review E*, 65:036308, 2002.

2
Delay Stabilization of Rotating Waves Without Odd Number Limitation

Bernold Fiedler, Valentin Flunkert, Marc Georgi, Philipp Hövel, and Eckehard Schöll

2.1
Introduction

A variety of methods have been developed in nonlinear science to stabilize unstable periodic orbits (UPOs) and control chaos [1], following the seminal work by Ott, Grebogi and Yorke [2], who employed a tiny control force to stabilize UPOs embedded in a chaotic attractor [3, 4]. A particularly simple and efficient scheme is time-delayed feedback as suggested by Pyragas [5], which uses the difference $z(t-\tau) - z(t)$ of a signal z at a time t and a delayed time $t-\tau$. It is an attempt to stabilize periodic orbits of (minimal) period T by a feedback control which involves a time delay $\tau = nT$, for suitable positive integer n. A linear feedback example is

$$\dot{z}(t) = f(\lambda, z(t)) + B[z(t-\tau) - z(t)], \tag{2.1}$$

where $\dot{z}(t) = f(\lambda, z(t))$ describes a d-dimensional nonlinear dynamical system with bifurcation parameter λ and an unstable orbit of (minimal) period T. The constant feedback control matrix B is chosen suitably. Typical choices are multiples of the identity or of rotations, or matrices of low rank. More general nonlinear feedbacks are conceivable, of course. The main point, however, is that the Pyragas choice $\tau_P = nT$ of the delay time eliminates the feedback term on the orbit, and thus recovers the original T-periodic solution $z(t)$. In this sense the method is noninvasive.

Although time-delayed feedback control has been widely used with great success in real world problems in physics, chemistry, biology, and medicine, e.g. [6–19], a severe limitation used to be imposed by the common belief that certain orbits cannot be stabilized for any strength of the control force. In fact, it had been contended that periodic orbits with an odd number of real Floquet multipliers greater than unity cannot be stabilized by the Pyragas method [20–25], even if the simple scheme (2.1) is extended by multiple delays in form of an infinite series [26]. To circumvent this restriction more complicated control schemes, like an oscillating feedback [27], half-period delays for

Review of Nonlinear Dynamics and Complexity. Edited by Heinz Georg Schuster

ISBN: 978-3-527-40729-3

special, symmetric orbits [28], or the introduction of an additional, unstable degree of freedom [25,29], have been proposed. Recently, however, it has been shown that the general limitation for orbits with an odd number of real unstable Floquet multipliers greater than unity does not hold: stabilization may be possible for suitable choices of B [30,31]. As an example, an unstable periodic orbit generated by a subcritical Hopf bifurcation has been considered. In particular, this refutes the theorem in [21]. In this article we review these recent findings, and give some general symmetry and stability considerations.

2.2
Mechanism of Stabilization

Consider the normal form of a subcritical Hopf bifurcation, extended by a time-delayed feedback term

$$\dot{z}(t) = \left[\lambda + i + (1 + i\gamma)|z(t)|^2\right] z(t) + b[z(t-\tau) - z(t)] \tag{2.2}$$

with $z \in \mathbb{C}$ and real parameters λ and γ. Here the Hopf frequency is normalized to unity. The feedback matrix B is represented by multiplication with a complex number $b = b_R + ib_I = b_0 e^{i\beta}$ with real b_R, b_I, β, and positive b_0. Note that the nonlinearity $f(\lambda, z(t)) = \left[\lambda + i + (1 + i\gamma)|z(t)|^2\right] z(t)$ and the control $B = b$ commute with complex rotations. Therefore $\exp(i\vartheta)z(t)$ solves Eq. (2.2), for any fixed ϑ, whenever $z(t)$ does. In particular, nonresonant Hopf bifurcations from the trivial solution $z \equiv 0$ at simple imaginary eigenvalues $\eta = i\omega \neq 0$ produce rotating wave solutions $z(t) = z(0)\exp\left(i\frac{2\pi}{T}t\right)$ with minimal period $T = 2\pi/\omega$ even in the nonlinear case and with delay terms. This follows from uniqueness of the emanating Hopf branches [32].

Transforming Eq. (2.2) to amplitude and phase variables $r(t), \varphi(t)$ for $z(t) = r(t)e^{i\varphi(t)}$, we obtain at vanishing control $b = 0$

$$\dot{r} = \left(\lambda + r^2\right) r \tag{2.3}$$

$$\dot{\varphi} = 1 + \gamma r^2. \tag{2.4}$$

An unstable periodic orbit (UPO) with $r = \sqrt{-\lambda}$ and period $T = 2\pi/(1 - \gamma\lambda)$ exists for $\lambda < 0$. This is the orbit which we will stabilize, called the *Pyragas orbit*. At $\lambda = 0$ a subcritical Hopf bifurcation occurs, and the steady state $z = 0$ loses its stability. The Pyragas control method is a noninvasive method: the control force vanishes at successful stabilization, and the periodic Pyragas orbit itself remains untouched by the control procedure. In order to satisfy this requirement the delays τ have to be chosen as a multiple of the minimal period: $\tau = nT$. This defines the local *Pyragas curve* in the (λ, τ)-plane for any $n \in \mathbb{N}$

$$\tau_P(\lambda) = \frac{2\pi n}{1 - \gamma\lambda} = 2\pi n(1 + \gamma\lambda + \dots), \tag{2.5}$$

emanating from the Hopf bifurcation points $\lambda = 0,\ \tau = 2\pi n$, towards negative λ.

Under further nondegeneracy conditions, the Hopf point $\lambda = 0,\ \tau = nT$ ($n \in \mathbb{N}_0$) also continues to a Hopf bifurcation curve $\tau = \tau_H(\lambda)$ for $\lambda < 0$. We determine this *Hopf curve* next. It is characterized by purely imaginary eigenvalues $\eta = i\omega$ of the transcendental characteristic equation

$$\eta = \lambda + i + b\left(e^{-\eta\tau} - 1\right) \tag{2.6}$$

which results from the linearization at the steady state $z = 0$ of the delayed system Eq. (2.2). Separating Eq. (2.6) into real and imaginary parts

$$0 = \lambda + b_0[\cos(\beta - \omega\tau) - \cos\beta] \tag{2.7}$$

$$\omega - 1 = b_0[\sin(\beta - \omega\tau) - \sin\beta] \tag{2.8}$$

and using the trigonometric identity

$$[\cos(\beta - \omega\tau)]^2 + [\sin(\beta - \omega\tau)]^2 = 1 \tag{2.9}$$

to eliminate $\omega(\lambda)$ from Eqs. (2.7) and (2.8) yields an explicit expression for the multivalued Hopf curve $\tau_H(\lambda)$ for given control amplitude b_0 and phase β:

$$\tau_H = \frac{\pm\arccos\left(\frac{b_0\cos\beta - \lambda}{b_0}\right) + \beta + 2\pi n}{1 - b_0\sin\beta \mp \sqrt{\lambda(2b_0\cos\beta - \lambda) + b_0^2\sin^2\beta}}. \tag{2.10}$$

Note that τ_H is not defined in case of $\beta = 0$ and $\lambda < 0$. Thus complex b is a necessary condition for the existence of the Hopf curve in the subcritical regime $\lambda < 0$. Figure 2.1 jointly displays the family of Hopf curves $n = 0, 1, \dots$ (solid), Eq. (2.10), and the Pyragas curve $n = 1$ (dashed), Eq. (2.5), in the (λ, τ) plane. In Fig. 2.1(b) the domains of instability of the trivial steady state $z = 0$, bounded by the Hopf curves, are marked by light gray shading. The dimensions of the unstable manifold of $z = 0$ are given in parentheses along the τ-axis in Fig. 2.1(b). By construction, the delay τ becomes a multiple of the minimal period T of the bifurcating Pyragas orbits along the Pyragas curve $\tau = \tau_p(\lambda) = nT$, and the time-delayed feedback term vanishes on these periodic orbits, noninvasively.

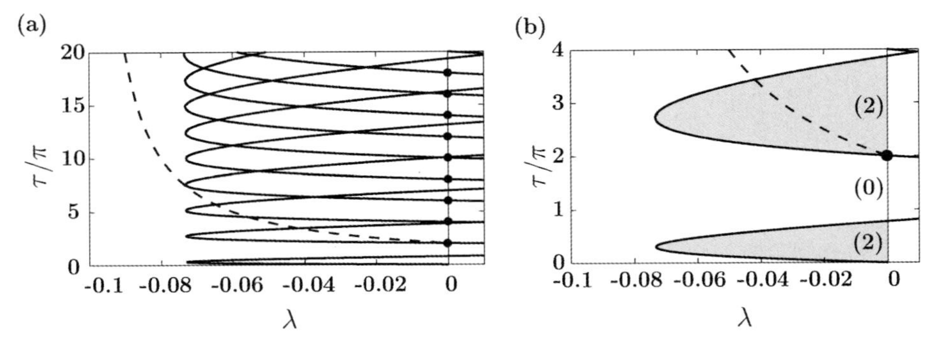

Fig. 2.1 Pyragas (dashed) and Hopf (solid) curves in the (λ, τ)-plane: (a) Hopf bifurcation curves $n = 0, \ldots, 10$, (b) Hopf bifurcation curves $n = 0, 1$ in an enlarged scale. Light gray shading marks the domains of unstable $z = 0$ and numbers in parentheses denote the dimension of the unstable manifold of $z = 0$ ($\gamma = -10$, $b_0 = 0.3$ and $\beta = \pi/4$).

Standard exchange of stability results [32], which hold verbatim for delay equations, then assert that the bifurcating branch of periodic solutions locally inherits linear asymptotic (in)stability from the trivial steady state, i.e., it consists of stable periodic orbits on the Pyragas curve $\tau_P(\lambda)$ inside the shaded domains for small $|\lambda|$. We stress that an unstable trivial steady state is not a sufficient condition for stabilization of the Pyragas orbit. In fact, the stabilized Pyragas orbit can destabilize again when $\lambda < 0$ is decreased further. This may be caused, for instance, by a torus bifurcation, see Section 2.5. However, there exists an interval of $\lambda < 0$ in our example for which Pyragas stabilization holds.

More precisely, periodic orbits of the controlled and uncontrolled system, alike, possess a unique and simple nontrivial Floquet multiplier $\mu = \exp(\Lambda T)$, near unity, as long as $|\lambda|$ remains small. All other nontrivial Floquet multipliers lie strictly inside the complex unit circle. In particular, the (strong) unstable dimension of these periodic orbits is either 0 or 1, depending on $\mu < 1$ or $\mu > 1$, respectively, and their unstable manifold is absent or two-dimensional. This is shown in Fig. 2.2 panel (a) top, which depicts solutions Λ of the characteristic equation of the periodic solution on the Pyragas curve (see Appendix).

The largest real part is positive for $b_0 = 0$. Thus the periodic orbit of the uncontrolled system is born unstable. As the amplitude of the feedback gain increases, the largest real part of the eigenvalue becomes smaller and eventually changes sign at TC. Hence the periodic orbit is stabilized. Note that an infinite number of Floquet exponents are created by the control scheme; their real parts tend to $-\infty$ in the limit $b_0 \to 0$ of vanishing control. Some of them may cross over to positive real parts for larger b_0 (dashed line in Fig. 2.2a), terminating the stability of the periodic orbit; see Section 2.5. Panel (a) bottom

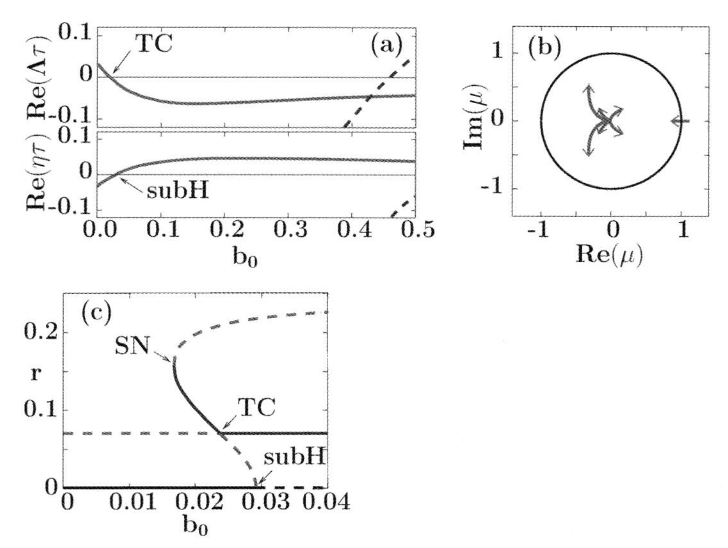

Fig. 2.2 (a) Top: Real part of Floquet exponents Λ of the periodic orbit vs. feedback amplitude b_0. Bottom: Real part of eigenvalue η of steady state vs. feedback amplitude b_0. (b) Floquet multipliers $\mu = \exp(\Lambda T)$ in the complex plane with the feedback amplitude $b_0 \in [0, 0.3]$ as a parameter. (c) Radii of periodic orbits. Solid (dashed) lines correspond to stable (unstable) orbits. ($\lambda = -0.005$, $\gamma = -10$, $\tau = \frac{2\pi}{1-\gamma\lambda}$, $\beta = \pi/4$).

illustrates the stability of the steady state by displaying the largest real part of the eigenvalues η.

Figure 2.2 (b) shows the behavior of the Floquet multipliers $\mu = \exp(\Lambda T)$ of the Pyragas orbit in the complex plane with increasing amplitude of the feedback gain b_0 as a parameter (marked by arrows). There is an isolated real multiplier crossing the unit circle at $\mu = 1$, in contrast to the result stated in [21]. This is caused by a transcritical bifurcation at which the Pyragas orbit collides with a delay-induced stable periodic orbit. In panel (c) of Fig. 2.2 the radii of all circular periodic orbits ($r = const$) are plotted versus the feedback strength b_0. For small b_0 only the initial (unstable) Pyragas orbit (T and r independent of b_0) and the steady state $r = 0$ (stable) exist. With increasing b_0 a pair of unstable/stable periodic orbits is created in a saddle-node (SN) bifurcation. The stable one of the two orbits (solid) then exchanges stability with the Pyragas orbit in a transcritical bifurcation (TC), and terminates at a subcritical Hopf bifurcation (subH), where the steady state $r = 0$ becomes unstable. The Pyragas orbit continues as a stable periodic orbit for larger b_0. Except at TC, the delay-induced orbit has a period $T \neq \tau$. Note that the respective exchanges of stability of the Pyragas orbit (TC) and the steady state (subH) occur at slightly different values of b_0. This is also corroborated by Fig. 2.2(c). The mechanism of stabilization of the Pyragas orbit by a transcritical bifurcation relies upon the possible existence of such delay-induced periodic orbits with $T \neq \tau$. Technically, the proof of the odd-number limitation theorem in [21] fails because the trivial Floquet multiplier $\mu = 1$ (Goldstone mode of periodic orbit) was neglected there; $F(1)$ in Eq. (14) in [21] is thus zero and not less than zero, as assumed [33]. At TC, where a second Floquet multiplier crosses the unit circle, this results in a Floquet multiplier $\mu = 1$ of algebraic multiplicity two.

2.3
S^1-Symmetry and Stability of Rotating Waves

For any fixed real ω, the transformation $z(t) = e^{i\omega t}\zeta(t)$ to co-rotating complex coordinates $\zeta(t)$ transforms Eq. (2.2) into the equivalent delay equation

$$\begin{aligned}\dot{\zeta}(t) &= (\lambda + (1-\omega)i)\zeta(t) + (1+i\gamma)|\zeta(t)|^2\zeta(t) \\ &\quad - b_0 \exp(i\beta)\left(\zeta(t) - e^{-i\omega\tau}\zeta(t-\tau)\right) .\end{aligned} \tag{2.11}$$

The co-rotating equation (2.11) remains autonomous since $e^{i\vartheta}z(t)$ solves Eq. (2.2) for any fixed real ϑ, whenever $z(t)$ does. Steady states $\dot{\zeta} = 0$ of Eq. (2.11) are precisely the rotating waves of Eq. (2.2), i.e., solutions of the form $z(t) = e^{i\omega t}\zeta_0$ with nonzero rotation frequency ω and nonzero $\zeta_0 \in \mathbb{C}$. The minimal period of such solutions $z(t)$ is of course given by $T = 2\pi/|\omega|$.

For $\tau = nT = 2\pi n/|\omega|$ with $n = 1, 2, 3, \ldots$ we obtain the non-invasive Pyragas control and the Pyragas curves (2.5).

Rotating waves, alias nontrivial steady states $\zeta(t) = \zeta_0$ of Eq. (2.11), are solutions of

$$0 = \lambda + (1-\omega)i + (1+i\gamma)|\zeta_0|^2 - b_0 \exp(i\beta)\left(1 - e^{-i\omega\tau}\right) . \tag{2.12}$$

The bifurcation diagram in Fig 2.2 (c) displays solutions (ω, r), $r = |\zeta_0|$, of Eq. (2.12) only. We can solve this equation for real parts

$$|\zeta_0|^2 = -\lambda - b_0[\cos(\beta - \omega\tau) - \cos(\beta)] \tag{2.13}$$

under the constraint of a positive right hand side. Substituting into the imaginary part of Eq. (2.12) yields the real equation

$$\begin{aligned}0 &= 1 - \omega + b_0[\sin(\beta - \omega\tau) - \sin(\beta)] \\ &\quad - \gamma\{\lambda + b_0[\cos(\beta - \omega\tau) - \cos(\beta)]\}.\end{aligned} \tag{2.14}$$

We seek solutions ω, depending on the five real parameters $\lambda, \gamma, b_0, \beta, \tau$. The degenerate case $\omega = 0$ which corresponds to a circle of equilibria, alias a frozen wave of vanishing angular velocity ω, arises only for $\gamma\lambda = 1$ [34], i.e., far away from the Hopf bifurcation.

The stabilization mechanism outlined in Section 2.2 can be understood from a mathematical view point as follows. At Hopf bifurcation we have a simple pair of purely imaginary eigenvalues, and no other imaginary eigenvalues. Therefore, the center manifold is two-dimensional at Hopf bifurcation [32]. Dimension two extends and includes the nearby transcritical bifurcation of rotating waves. Moreover, the center manifold can be chosen to be invariant with respect to the S^1 action $z \mapsto e^{i\vartheta} z$ [35]. In polar coordinates $z = r\, e^{i\varphi}$ the dynamics in any two-dimensional center manifold is therefore given by a system of the general form

$$\dot{r} = f(r^2, \underline{\mu})r \tag{2.15a}$$

$$\dot{\varphi} = g(r^2, \underline{\mu}) \tag{2.15b}$$

with parameter vector $\underline{\mu}$, i.e., in our case $\underline{\mu} = (\lambda, \gamma, b_0, \beta, \tau)$. Note that φ neither enters the equation for $\dot{r}$ nor for $\dot{\varphi}$. Indeed, $(r(t), \varphi(t) + \vartheta)$ must be a solution for any fixed ϑ, by S^1-equivariance, whenever $(r(t), \varphi(t))$ is. Also note that Eq. (2.15a) is a system of differential equations which does not involve time-delayed arguments. Rather, the original time delay τ enters as one parameter among others.

To determine f on the center manifold, we first observe that $f(r^2, \underline{\mu}) = 0$ defines rotating (or frozen) waves with $|\zeta_0| = r$, and thus must be equivalent to Eq. (2.12) with $\omega = g(r^2, \underline{\mu})$. Conversely, the solution set $(r^2, \omega, \underline{\mu})$ is therefore given by Eqs. (2.13) and (2.14), and defines the zero set of f. Again, $f(r^2, \underline{\mu}) = 0$ if, and only if, $(\omega, \underline{\mu})$ solve Eq. (2.14) and r^2 is given by Eq. (2.13).

To determine the stability of our rotating waves within the center manifold it remains to determine the sign of f outside the zero set. That sign is known at the trivial equilibrium $r = 0$, by standard exchange of stability at nondegenerate Hopf bifurcations. Normally hyperbolic rotating waves correspond to simple zeros of f in the r-direction, i.e., $\partial_r f \neq 0$. This allows us to determine the sign of f in the bifurcation diagram Fig. 2.2 (c). The (in-)stability properties of all rotating waves within the two-dimensional center manifold are then immediate. Spectral analysis at the Hopf bifurcation shows strict stability of the remaining nonimaginary eigenvalues. Therefore the center manifold is attractive [32]. In particular, (in-)stability of rotating waves in the full delay system (2.2) is inherited from the center manifold analysis without ever computing the manifold itself. In conclusion this proves the stability properties indicated in Fig. 2.2 (c), for the parameters chosen there.

2.4 Conditions on the Feedback Gain

Next we analyse the conditions under which stabilization of the subcritical periodic orbit is possible. From Fig. 2.1(b) it is evident that the Pyragas curve must lie inside the light gray region, i.e., the Pyragas and Hopf curves emanating from the point $(\lambda, \tau) = (0, 2\pi)$ must locally satisfy the inequality $\tau_H(\lambda) < \tau_P(\lambda)$ for $\lambda < 0$. More generally, let us investigate the eigenvalue crossings of the Hopf eigenvalues $\eta = i\omega$ along the τ-axis of Fig. 2.1. In particular we derive conditions for the unstable dimensions of the trivial steady state near the Hopf bifurcation point $\lambda = 0$ in our model equation (2.2). On the τ-axis ($\lambda = 0$), the characteristic equation (2.6) for $\eta = i\omega$ is reduced to

$$\eta = i + b\left(e^{-\eta\tau} - 1\right), \tag{2.16}$$

and we obtain two series of Hopf points given by

$$0 \leq \tau_n^A = 2\pi n \tag{2.17}$$

$$0 < \tau_n^B = \frac{2\beta + 2\pi n}{1 - 2b_0 \sin\beta} \quad (n = 0, 1, 2, \ldots). \tag{2.18}$$

The corresponding Hopf frequencies are $\omega^A = 1$ and $\omega^B = 1 - 2b_0 \sin\beta$, respectively. Note that series A consists of all Pyragas points, since $\tau_n^A = nT = \frac{2\pi n}{\omega^A}$. In the series B the integers n have to be chosen such that the delay $\tau_n^B \geq 0$. The case $b_0 \sin\beta = 1/2$, only, corresponds to $\omega^B = 0$ and does not occur for finite delays τ.

We evaluate the crossing directions of the critical Hopf eigenvalues next, along the positive τ-axis and for both series. Abbreviating $\frac{\partial}{\partial\tau}\eta$ by η_τ the crossing direction is given by $\operatorname{sign}(\operatorname{Re}\eta_\tau)$. Implicit differentiation of Eq. (2.16) with respect to τ at $\eta = i\omega$ implies

$$\operatorname{sign}(\operatorname{Re}\eta_\tau) = -\operatorname{sign}(\omega)\operatorname{sign}(\sin(\omega\tau - \beta)). \tag{2.19}$$

We are interested specifically in the Pyragas-Hopf points of series A (marked by dots in Fig. 2.1) where $\tau = \tau_n^A = 2\pi n$ and $\omega = \omega^A = 1$. Indeed $\operatorname{sign}(\operatorname{Re}\eta_\tau) = \operatorname{sign}(\sin\beta) > 0$ holds, provided we assume $0 < \beta < \pi$, i.e., $b_I > 0$ for the feedback gain. This condition alone, however, is not sufficient to guarantee stability of the steady state for $\tau < 2n\pi$. We also have to consider the crossing direction $\operatorname{sign}(\operatorname{Re}\eta_\tau)$ along series B, $\omega^B = 1 - 2b_0 \sin\beta$, $\omega^B \tau_n^B = 2\beta + 2\pi n$, for $0 < \beta < \pi$. Equation (2.19) now implies $\operatorname{sign}(\operatorname{Re}\eta_\tau) = \operatorname{sign}((2b_0 \sin\beta - 1)\sin\beta) = \operatorname{sign}(2b_0 \sin\beta - 1)$.

To compensate for the destabilization of $z = 0$ upon each crossing of any point $\tau_n^A = 2\pi n$, we must require stabilization ($\operatorname{sign}(\operatorname{Re}\eta_\tau) < 0$) at each point

τ_n^B of series B. If $b_0 \geq 1/2$, this requires $0 < \beta < \arcsin(1/(2b_0))$ or $\pi - \arcsin(1/(2b_0)) < \beta < \pi$. The distance between two successive points τ_n^B and τ_{n+1}^B is $2\pi/\omega^B > 2\pi$. Therefore, there is at most one τ_n^B between any two successive Hopf points of series A. Stabilization requires exactly one such τ_n^B, specifically: $\tau_{k-1}^A < \tau_{k-1}^B < \tau_k^A$ for all $k = 1, 2, \ldots, n$. This condition is satisfied if, and only if,

$$0 < \beta < \beta_n^*, \tag{2.20}$$

where $0 < \beta_n^* < \pi$ is the unique solution of the transcendental equation

$$\frac{1}{\pi}\beta_n^* + 2nb_0 \sin\beta_n^* = 1. \tag{2.21}$$

This holds because the condition $\tau_{k-1}^A < \tau_{k-1}^B < \tau_k^A$ first fails when $\tau_{k-1}^B = \tau_k^A$. Equation (2.20) represents a necessary but not yet sufficient condition that the Pyragas choice $\tau_P = nT$ for the delay time will stabilize the periodic orbit.

To evaluate the remaining condition, $\tau_H < \tau_P$ near $(\lambda, \tau) = (0, 2\pi)$, we expand the exponential in the characteristic equation (2.6) for $\omega\tau \approx 2\pi n$, and obtain the approximate Hopf curve for small $|\lambda|$:

$$\tau_H(\lambda) \approx 2\pi n - \frac{1}{b_I}(2\pi n b_R + 1)\lambda. \tag{2.22}$$

Recalling Eq. (2.5), the Pyragas stabilization condition $\tau_H(\lambda) < \tau_P(\lambda)$ is therefore satisfied for $\lambda < 0$ if, and only if,

$$\frac{1}{b_I}\left(b_R + \frac{1}{2\pi n}\right) < -\gamma. \tag{2.23}$$

Equation (2.23) defines a domain in the plane of the complex feedback gain $b = b_R + ib_I = b_0 e^{i\beta}$ bounded from below (for $\gamma < 0 < b_I$) by the straight line

$$b_I = \frac{1}{-\gamma}\left(b_R + \frac{1}{2\pi n}\right). \tag{2.24}$$

Equation (2.21) represents a curve $b_0(\beta)$, i.e.,

$$b_0 = \frac{1}{2n\sin\beta}\left(1 - \frac{\beta}{\pi}\right), \tag{2.25}$$

which forms the upper boundary of a domain given by the inequality (2.20). Thus Eqs. (2.24) and (2.25) describe the boundaries of the domain of control in the complex plane of the feedback gain b in the limit of small λ. Figure 2.3 depicts this domain of control for $n = 1$, i.e., a time delay $\tau = \frac{2\pi}{1-\gamma\lambda}$. The lower and upper solid curves correspond to Eqs. (2.24) and (2.25), respectively. The grayscale displays the numerical result of the largest real part of the Floquet exponents, wherever Re $\Lambda < 0$, calculated from linearization of the amplitude and phase equations around the periodic orbit (see Appendix). Outside the shaded areas the periodic orbit is not stabilized. With increasing $|\lambda|$ the domain of stabilization shrinks, as the deviations from the linear approximation (2.22) become larger. For sufficiently large $|\lambda|$ stabilization is no longer possible, in agreement with Fig. 2.1(b). Note that for real values of b, i.e., $\beta = 0$, no stabilization occurs at all. Hence, stabilization fails if the feedback matrix B is a multiple of the identity matrix.

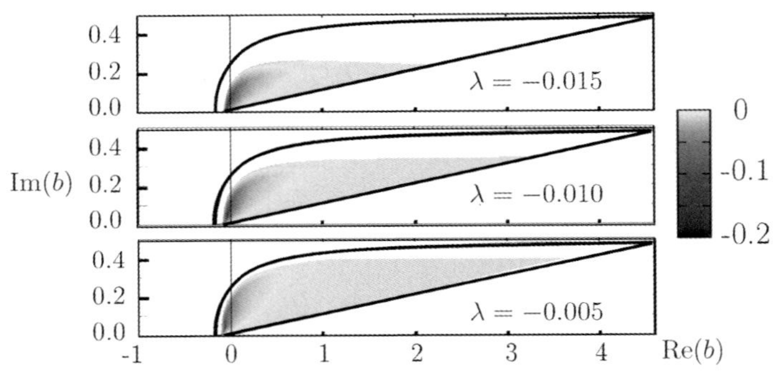

Fig. 2.3 Domain of control in the plane of the complex feedback gain $b = b_0 e^{i\beta}$ for three different values of the bifurcation parameter λ. The solid curves indicate the boundary of stability in the limit $\lambda \nearrow 0$, see Eqs. (2.24) and (2.25). The shading shows the magnitude of the largest (negative) real part of the Floquet exponents of the periodic orbit ($\gamma = -10$, $\tau = \frac{2\pi}{1-\gamma\lambda}$).

2.5 Tori

Rotating waves $z(t) = e^{i\omega t}\zeta_0$ are periodic solutions of Eq. (2.2). But not all periodic solutions need to be rotating waves. Any bifurcation from rotating waves, however, must be visible in co-rotating coordinates $\zeta(t) = e^{-i\omega t}z(t)$ as well. Since rotating waves of Eq. (2.2) are equilibria of Eq. (2.11) any such bifurcation must be accompanied by purely imaginary eigenvalues Λ of the characteristic equation associated to ζ_0.

At the rotating wave equilibrium $r_0 = \zeta_0 > 0$ of the co-rotating system Eq. (2.11), the characteristic equation for Floquet multipliers $\mu = e^{\Lambda T}$ becomes

$$0 = \chi(\Lambda) = \Lambda^2 - 2({r_0}^2 + b_0 cE)\Lambda + 2{r_0}^2(c + \gamma s)b_0 E + (b_0 E)^2.$$

See the appendix for a derivation. Here we have abbreviated $c = \cos\beta$, $s = \sin\beta$ for the control $b = b_0 e^{\beta}$, and $E = e^{-\Lambda\tau} - 1$ for the Floquet exponents Λ.

In Fig. 2.2 (a) we have indicated how the real part $\mathrm{Re}\Lambda$ of a delay induced Floquet exponent of the Pyragas orbit $z(t)$ crosses zero when the control amplitude b_0 is increased (dashed line). In Figs. 2.4 and 2.5 we indicate the resulting dynamic consequences for a particular choice of parameters.

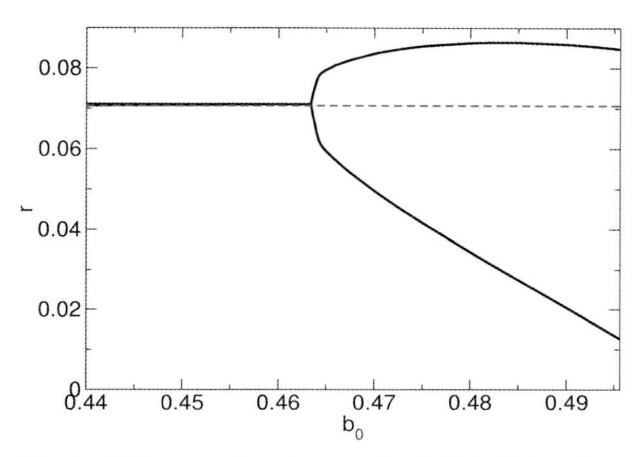

Fig. 2.4 Minimum and maximum radius (solid), obtained from simulations, and radius of the Pyragas orbit $r = \sqrt{-\lambda}$ (dashed line) versus the control amplitude b_0. For large b_0 the stabilized Pyragas orbit becomes unstable in a torus (Neimark–Sacker) bifurcation (compare Fig. 2.2 (a), dashed line) $(\gamma = -10, \lambda = -0.005, \beta = \pi/4, \tau = \frac{2\pi}{1-\gamma\lambda})$.

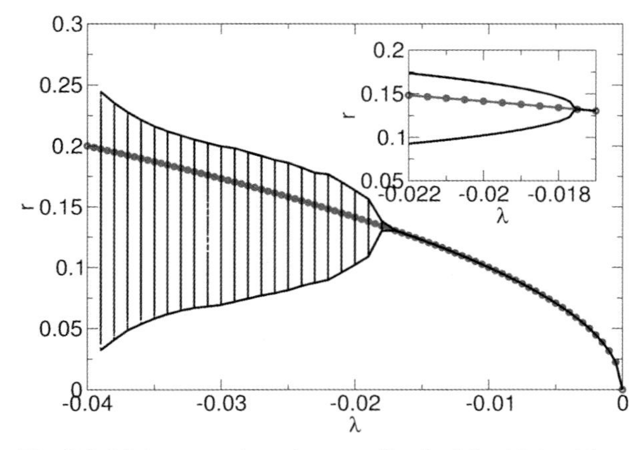

Fig. 2.5 Minimum and maximum radius (solid), obtained from simulations, and radius of the Pyragas orbit $r = \sqrt{-\lambda}$ (circles) versus λ. For sufficiently negative λ the stabilized Pyragas orbit becomes unstable in a torus Neimark–Sacker bifurcation $(\gamma = -10, b_0 = 0.3, \beta = \pi/4, \tau = \frac{2\pi}{1-\gamma\lambda})$.

The observed amplitude oscillations can be explained by a supercritical Neimark–Sacker torus bifurcation scenario. Indeed the nontrivial purely imaginary Floquet exponent Λ causes the associated nonreal Floquet multiplier $\mu = e^{\Lambda T}$ to cross the complex unit circle and destabilize the periodic Pyragas orbit inspite of an increasing control amplitude (see Fig. 2.4). The resulting bifurcation is supercritical in our example; see also Fig. 2.5. The bifurcating 2-torus inherits stability, taking it away from the Pyragas orbit by exchange of stability.

Equivariance under rotations causes a slight improvement over standard torus bifurcation. The rotation number along the bifurcating torus branch, which would behave in a devil's staircase manner in the general case, in fact becomes a smooth function of the amplitude of the resulting oscillations of $r(t) = |z(t)|$. In particular, the usual phase locking or resonance phenomena at rational rotation numbers on invariant tori do not occur. This can be seen, for example, by an analysis in suitable Palais coordinates along the relative equilibrium to the S^1-action which is given by the destabilized Pyragas orbit. Eliminating the S^1-action, the Neimark–Sacker bifurcation then becomes a nondegenerate relative Hopf bifurcation from the relative equilibrium $\zeta_0 = r_0$. The Hopf eigenvalues are provided by the purely imaginary Floquet exponent Λ. See Ref. [36] and the survey [37] for details of such a setting. A similar phenomenon occurs, albeit in a reaction-diffusion setting, when rigidly rotating spiral waves in excitable media destabilize to meandering motions [37,38]. For the likewise related destabilization of wavy vortices to modulated wavy vortices in Taylor-Couette fluid flow between rotating concentric cylinders see [39] and the references there. Intensity pulsations in laser systems provide yet another experimental source of the same phenomenon of smooth parameter dependence of rotation numbers (see Refs. [40,41]).

Unlike the present case, though, these phenomena do not involve delayed feedback control. In our case of Pyragas control we can conclude that stabilization and destabilization of Pyragas orbits occurs, either by transcritical bifurcations of non-Pyragas periodic orbits, or else by Neimark–Sacker torus bifurcation with nonresonant, smooth dependence of the rotation number.

2.6 Conclusion

In conclusion, we have refuted the claim that a periodic orbit with an odd number of real Floquet multipliers greater than unity cannot be stabilized by time-delayed feedback control of Pyragas type. For this purpose we have analyzed the generic example of the normal form of a subcritical Hopf bifurcation, which is paradigmatic for a large class of nonlinear systems. We have worked out explicit analytical conditions for stabilization of the periodic orbit generated by a subcritical Hopf bifurcation in terms of the amplitude and the phase of the feedback control gain. Our results emphasize the crucial role of a non-vanishing phase of the control signal for stabilization of periodic orbits which violate the odd-number limitation. The feedback phase is readily accessible and can be adjusted, for instance, in laser systems, where subcritical Hopf bifurcation scenarios are abundant and Pyragas control can be realized via coupling to an external Fabry-Perot resonator [18]. The importance of the feedback phase for the stabilization of steady states in lasers [18] and neural systems [42], as well as for stabilization of periodic orbits by a time-delayed feedback control scheme using spatio-temporal filtering [43], has been noted recently. Here, we have shown that the odd-number limitation does not hold in general, which opens up fundamental questions as well as a wide range of applications. The result will not only be important for practical applications in physical sciences, technology, and life sciences, where one might often desire to stabilize periodic orbits with an odd number of positive Floquet exponents. It also suggests noninvasive experimental and numerical tracking of unstable orbits and bifurcation analysis using time-delayed feedback control [44].

Acknowledgements

This work was supported by Deutsche Forschungsgemeinschaft in the framework of Sfb 555. We acknowledge helpful discussions with A. Amann, W. Just, and A. Pikovsky.

Appendix: Calculation of Floquet Exponents

The Floquet exponents Λ of the Pyragas orbit can be calculated explicitly by rewriting Eq. (2.2) in polar coordinates $z(t) = r(t)\, e^{i\varphi(t)}$

$$\dot{r}(t) = (\lambda + r^2)\, r + b_0[\cos(\beta + \varphi(t-\tau) - \varphi)\, r(t-\tau) - r\, \cos\beta] \tag{2.26}$$

$$\dot{\varphi}(t) = 1 + \gamma r^2 + b_0[\sin(\beta + \varphi(t-\tau) - \varphi)\, \frac{r(t-\tau)}{r} - \sin\beta] \tag{2.27}$$

and linearizing around the periodic orbit according to $r(t) = r_0 + \delta r(t)$ and $\varphi(t) = \Omega t + \delta\varphi(t)$, with $r_0 = \sqrt{-\lambda}$ and $\Omega = 1 - \gamma\lambda$ (see Eq. (2.3)). This yields

$$\begin{aligned} \begin{pmatrix} \dot{\delta r}(t) \\ \dot{\delta\varphi}(t) \end{pmatrix} &= \begin{bmatrix} -2\lambda - b_0\cos\beta & b_0 r_0 \sin\beta \\ 2\gamma r_0 - b_0 \sin\beta\, r_0^{-1} & -b_0\cos\beta \end{bmatrix} \begin{pmatrix} \delta r(t) \\ \delta\varphi(t) \end{pmatrix} \\ &\quad + \begin{bmatrix} b_0\cos\beta & -b_0 r_0 \sin\beta \\ b_0 \sin\beta r_0^{-1} & b_0\cos\beta \end{bmatrix} \begin{pmatrix} \delta r(t-\tau) \\ \delta\varphi(t-\tau) \end{pmatrix}. \end{aligned}$$

With the ansatz

$$\begin{pmatrix} \delta r(t) \\ \delta\varphi(t) \end{pmatrix} = u\ \exp(\Lambda t), \tag{2.28}$$

where u is a two-dimensional vector, one obtains the autonomous linear equation

$$\begin{bmatrix} -2\lambda + b_0\cos\beta\,(e^{-\Lambda\tau} - 1) - \Lambda & -b_0 r_0 \sin\beta\,(e^{-\Lambda\tau} - 1) \\ 2\gamma r_0 + b_0 r_0^{-1} \sin\beta\,(e^{-\Lambda\tau} - 1) & b_0\cos\beta\,(e^{-\Lambda\tau} - 1) - \Lambda \end{bmatrix} u = 0. \tag{2.29}$$

The condition of vanishing determinant then gives the transcendental characteristic equation

$$\begin{aligned} 0 &= \left(-2\lambda + b_0\cos\beta\,(e^{-\Lambda\tau} - 1) - \Lambda\right) \left(b_0\cos\beta\,(e^{-\Lambda\tau} - 1) - \Lambda\right) \\ &\quad + b_0 r_0 \sin\beta\,(e^{-\Lambda\tau} - 1) \left(2\gamma r_0 + b_0 r_0^{-1} \sin\beta\,(e^{-\Lambda\tau} - 1)\right) \end{aligned} \tag{2.30}$$

for the Floquet exponents Λ which can be solved numerically.

References

1 E. Schöll and H. G. Schuster (Eds.): *Handbook of Chaos Control*, second completely revised and enlarged edition. Wiley-VCH, Weinheim, 2007.

2 E. Ott, C. Grebogi, and J. A. Yorke: *Controlling chaos*, Phys. Rev. Lett. **64**, 1196 (1990).

3 S. Boccaletti, C. Grebogi, Y. C. Lai, H. Mancini, and D. Maza: *The control of chaos: theory and applications*, Phys. Rep. **329**, 103 (2000).

4 D. J. Gauthier: *Resource letter: Controlling chaos*, Am. J. Phys. **71**, 750 (2003).

5 K. Pyragas: *Continuous control of chaos by self-controlling feedback*, Phys. Lett. A **170**, 421 (1992).

6 K. Pyragas and A. Tamaševičius: *Experimental control of chaos by delayed self-controlling feedback*, Phys. Lett. A **180**, 99 (1993).

7 S. Bielawski, D. Derozier, and P. Glorieux: *Controlling unstable periodic orbits by a delayed continuous feedback*, Phys. Rev. E **49**, R971 (1994).

8 T. Pierre, G. Bonhomme, and A. Atipo: *Controlling the chaotic regime of nonlinear ionization waves using time-delay autosynchronisation method*, Phys. Rev. Lett. **76**, 2290 (1996).

9 K. Hall, D. J. Christini, M. Tremblay, J. J. Collins, L. Glass, and J. Billette: *Dynamic control of cardiac alterans*, Phys. Rev. Lett. **78**, 4518 (1997).

10 D. W. Sukow, M. E. Bleich, D. J. Gauthier, and J. E. S. Socolar: *Controlling chaos in a fast diode resonator using time-delay autosynchronisation: Experimental observations and theoretical analysis*, Chaos **7**, 560 (1997).

11 O. Lüthje, S. Wolff, and G. Pfister: *Control of chaotic taylor-couette flow with time-delayed feedback*, Phys. Rev. Lett. **86**, 1745 (2001).

12 P. Parmananda, R. Madrigal, M. Rivera, L. Nyikos, I. Z. Kiss, and V. Gáspár: *Stabilization of unstable steady states and periodic orbits in an electrochemical system using delayed-feedback control*, Phys. Rev. E **59**, 5266 (1999).

13 J. M. Krodkiewski and J. S. Faragher: *Stabilization of motion of helicopter rotor blades using delayed feedback – modelling, computer simulation and experimental verification*, J. Sound Vib. **234**, 591 (2000).

14 T. Fukuyama, H. Shirahama, and Y. Kawai: *Dynamical control of the chaotic state of the current-driven ion acoustic instability in a laboratory plasma using delayed feedback*, Phys. Plasmas **9**, 4525 (2002).

15 C. von Loewenich, H. Benner, and W. Just: *Experimental relevance of global properties of time-delayed feedback control*, Phys. Rev. Lett. **93**, 174101 (2004).

16 M. G. Rosenblum and A. Pikovsky: *Controlling synchronization in an ensemble of globally coupled oscillators*, Phys. Rev. Lett. **92**, 114102 (2004).

17 O. V. Popovych, C. Hauptmann, and P. A. Tass: *Effective desynchronization by nonlinear delayed feedback*, Phys. Rev. Lett. **94**, 164102 (2005).

18 S. Schikora, P. Hövel, H. J. Wünsche, E. Schöll, and F. Henneberger: *All-optical noninvasive control of unstable steady states in a semiconductor laser*, Phys. Rev. Lett. **97**, 213902 (2006).

19 E. Schöll, J. Hizanidis, P. Hövel, and G. Stegemann: *Pattern formation in semiconductors under the influence of time-delayed feedback control and noise*, in *Analysis and control of complex nonlinear processes in physics, chemistry and biology*, edited by L. Schimansky-Geier, B. Fiedler, J. Kurths, and E. Schöll. World Scientific, Singapore, 2007, pp. 135–183.

20 W. Just, T. Bernard, M. Ostheimer, E. Reibold, and H. Benner: *Mechanism of time-delayed feedback control*, Phys. Rev. Lett. **78**, 203 (1997).

21 H. Nakajima: *On analytical properties of delayed feedback control of chaos*, Phys. Lett. A **232**, 207 (1997).

22 H. Nakajima and Y. Ueda: *Limitation of generalized delayed feedback control*, Physica D **111**, 143 (1998).

23 I. Harrington and J. E. S. Socolar: *Limitation on stabilizing plane waves via time-delay feedback*, Phys. Rev. E **64**, 056206 (2001).

24 K. Pyragas, V. Pyragas, and H. Benner: *Delayed feedback control of dynamical systems at subcritical Hopf bifurcation*, Phys. Rev. E **70**, 056222 (2004).

25 V. Pyragas and K. Pyragas: *Delayed feedback control of the Lorenz system: An analytical treatment at a subcritical Hopf bifurcation*, Phys. Rev. E **73**, 036215 (2006).

26 J. E. S. Socolar, D. W. Sukow, and D. J. Gauthier: *Stabilizing unstable periodic orbits in fast dynamical systems*, Phys. Rev. E **50**, 3245 (1994).

27 H. G. Schuster and M. B. Stemmler: *Control of chaos by oscillating feedback*, Phys. Rev. E **56**, 6410 (1997).

28 H. Nakajima and Y. Ueda: *Half-period delayed feedback control for dynamical systems with symmetries*, Phys. Rev. E **58**, 1757 (1998).

29 K. Pyragas: *Control of chaos via an unstable delayed feedback controller*, Phys. Rev. Lett. **86**, 2265 (2001).

30 B. Fiedler, V. Flunkert, M. Georgi, P. Hövel, and E. Schöll: *Refuting the odd number limitation of time-delayed feedback control*, Phys. Rev. Lett. **98**, 114101 (2007).

31 W. Just, B. Fiedler, V. Flunkert, M. Georgi, P. Hövel, and E. Schöll: *Beyond odd number limitation: a bifurcation analysis of time-delayed feedback control*, Phys. Rev. E **76**, 02610 (2007).

32 O. Diekmann, S. A. van Gils, S. M. Verduyn Lunel, and H. O. Walther: *Delay Equations*. Springer-Verlag, New York, 1995.

33 A. Amann (private communication).

34 B. Fiedler: *Global Bifurcation of Periodic Solutions with Symmetry*. Springer-Verlag, Heidelberg, 1988.

35 A. Vanderbauwhede: *Centre manifolds, normal forms and elementary bifurcations*, Dynamics Reported **2**, 89 (1989).

36 B. Fiedler, B. Sandstede, A. Scheel, and C. Wulf: *Bifurcation from Relative Equilibria of Noncompact Group Actions: Skew Products, Meanders, and Drifts*, Documenta Mathematica **1**, 479 (1996).

37 B. Fiedler and A. Scheel: *Dynamics of reaction–diffusion patterns*, in *Trends in Nonlinear Analysis, Festschrift dedicated to Willi Jäger for his 60th birthday*, edited by M. Kirkilionis, R. Rannacher, and F. Tomi. Springer, Heidelberg, 2002.

38 B. Fiedler, M. Georgi, and N. Jangle: *Spiral wave dynamics: Reaction and diffusion versus kinematics*, in *Analysis and Control of complex nonlinear Processes in Physics, Chemistry and Biology*. World Scientific, Singapore, 2007.

39 M. Golubitsky and I. Stewart: *Singularities and Groups in Bifurcation Theory. Volume 2*, vol. 69. Springer-Verlag, New York, 1988.

40 S. Bauer, O. Brox, J. Kreissl, B. Sartorius, M. Radziunas, J. Sieber, H. J. Wünsche, and F. Henneberger: *Nonlinear dynamics of semiconductor lasers with active optical feedback*, Phys. Rev. E **69**, 016206 (2004).

41 S. Wieczorek, B. Krauskopf, and D. Lenstra: *Unifying view of bifurcations in a semiconductor laser subject to optical injection*, Optics Communications **172**, 279 (1999).

42 M. G. Rosenblum and A. Pikovsky: *Delayed feedback control of collective synchrony: An approach to suppression of pathological brain rhythms*, Phys. Rev. E **70**, 041904 (2004).

43 N. Baba, A. Amann, E. Schöll, and W. Just: *Giant improvement of time-delayed feedback control by spatio-temporal filtering*, Phys. Rev. Lett. **89**, 074101 (2002).

44 J. Sieber and B. Krauskopf: *Control based bifurcation analysis for experiments*, Nonlinear Dynamics, DOI: 10.1007/s11071-007-9217-2 (2007)

3
Random Boolean Networks

Barbara Drossel

3.1
Introduction

Random Boolean networks (RBNs) were introduced in 1969 by S. Kauffman [1, 2] as a simple model for gene regulatory networks. Each gene was represented by a node that has two possible states, "on" (corresponding to a gene that is being transcribed) and "off" (corresponding to a gene that is not being transcribed). There are altogether N nodes, and each node receives input from K randomly chosen nodes, which represent the genes that control the considered gene. Furthermore, each node is assigned an update function that prescribes the state of the node in the next time step, given the state of its input nodes. This update function is chosen from the set of all possible update functions according to some probability distribution. Starting from some initial configuration, the states of all nodes of the network are updated in parallel. Since configuration space is finite and since dynamics is deterministic, the system must eventually return to a configuration that it has had before, and from then on it repeats the same sequence of configurations periodically: it is on an *attractor*.

Kauffman focussed his interest on *critical* networks, which are at the boundary between *frozen* networks with only very short attractors and *chaotic* chaotic networks with attractors that may include a finite proportion of state space. He equated attractors with cell types. Since each cell contains the same DNA (i.e., the same network), cells can only differ by the pattern of gene activity. Based on results of computer simulations for the network sizes possible at that time, S. Kauffman found that the mean number of attractors in critical networks with $K = 2$ inputs per node increases as $\sqrt{N}$. This finding was very satisfying, since the biological data available at that time for various species indicated that the number of cell types is proportional to the square root of the number of genes. This would mean that the very simple model of RBNs with its random wiring and its random assignment of update functions displays

Review of Nonlinear Dynamics and Complexity. Edited by Heinz Georg Schuster

ISBN: 978-3-527-40729-3

the same scaling laws as the more complex reality. The concept of universality, familiar from equilibrium critical phenomena, appeared to work also for this class of nonequilibrium systems. Kauffman found also that the mean length of attractors increases as $\sqrt{N}$.

Today we know that the biological data and the computer simulation data are both incorrect. The sequencing of entire genomes in recent years revealed that the number of genes is not proportional to the mass of DNA (as was assumed at that time), but much smaller for higher organisms. The square-root law for attractor numbers and lengths in RBNs survived until RBNs were studied with much more powerful computers. Then it was found that for larger N the apparent square-root law does not hold any more, but that the increase with system size is faster. The numerical work was complemented by several beautiful analytical papers, and today we know that the attractor number and length of $K = 2$ networks increases with network size faster than any power law. We also know that, while attractor numbers do not obey power laws, other properties of critical RBNs do obey power laws.

It is the purpose of this review to explain in an understandable and self-contained way the properties of RBNs and their attractors, with a special focus on critical networks. To this aim, this review contains examples, short calculations, phenomenological arguments, and problems to solve. Long calculations and plots of computer simulation data were not included and are not necessary for the understanding of the arguments. The readers will also benefit from consulting the review [3], which, while not containing the more recent findings, covers many important topics related to Boolean networks.

Boolean networks are used not only to model gene regulation networks, but also neural networks, social networks, and protein interaction networks. The structure of all these networks is different from RBNs with their random wiring and random assignment of update functions, and with the same number of inputs for every node. Nevertheless, understanding RBNs is a first and important step on our way to understanding the more complex real networks.

3.2
Model

A random Boolean network is specified by its topology and its dynamical rules. The topology is given by the nodes and the links between these nodes. The links are directed, i.e., they have an arrow pointing from a node to those nodes that it influences. The dynamical rules describe how the states of the nodes change with time. The state of each node is "on" or "off", and it is determined by the state of the nodes that have links to it (i.e., that are its inputs). In the following, we first describe the topology, and then the dynamics of RBNs.

3.2.1 Topology

For a given number N of nodes and a given number K of inputs per node, a RBN is constructed by choosing the K inputs of each node at random among all nodes. If we construct a sufficiently large number of networks in this way, we generate an *ensemble* of networks. In this ensemble, all possible topologies occur, but their statistical weights are usually different. Let us consider the simplest possible example, $N = 2$ and $K = 1$, shown in Fig. 3.1. There are three possible topologies. Topologies (a) and (b) have each the statistical weight $1/4$ in the ensemble, since each of the links is connected in the given way with probability $1/2$. Topology (c) has the weight $1/2$, since there are two possibilities for realizing this topology: either of the two nodes can be the one with the self-link.

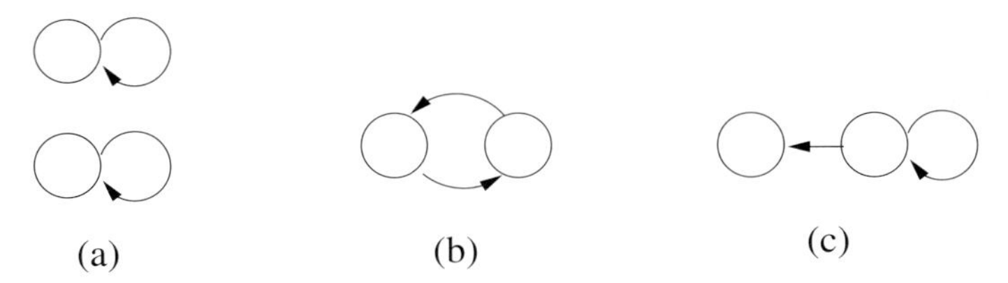

Fig. 3.1 The possible topologies for $N = 2$ and $K = 1$.

While the number of inputs of each node is fixed by the parameter K, the number of outputs (i.e. of outgoing links) varies between the nodes. The mean number of outputs must be K, since there must be in total the same number of outputs as inputs. A given node becomes the input of each of the N nodes with probability K/N. In the *thermodynamic limit* $N \to \infty$ the probability distribution of the number of outputs is therefore a Poisson distribution

$$P_{out}(k) = \frac{K^k}{k!} e^{-K} . \tag{3.1}$$

3.2.2 Update Functions

Next, let us specify the dynamics of the networks. Each node can be in the state $\sigma_i = 1$ ("on") or in the state $\sigma_i = 0$ ("off"), where i is the index of the node. The N nodes of the network can therefore together assume 2^N different states. An update function specifies the state of a node in the next time step, given the state of its K inputs at the present time step. Since each of the K inputs of a node can be on or off, there are $M = 2^K$ possible input states.

The update function has to specify the new state of a node for each of these input states. Consequently, there are 2^M different update functions.

Table 3.1 lists the 4 possible update functions for $K = 1$. The first two functions are constant, or "frozen", i.e. the state of the node is independent of its inputs. The other two functions change whenever an input changes, i.e., they are *reversible*. The third function is the "copy" function, the fourth is the "invert" function.

Table 3.1 The 4 update functions for nodes with 1 input. The first column lists the 2 possible states of the input, the other columns represent one update function each, falling into two classes.

In	$\mathcal{F}$		$\mathcal{R}$	
0	1	0	0	1
1	1	0	1	0

Table 3.2 lists the 16 possible update functions for $K = 2$. There are again two constant and two reversible functions. Furthermore, there are *canalyzing* functions. A function is canalyzing if at least for one value of one of its inputs the output is fixed, irrespective of the values of the other inputs. The first class of canalyzing functions do not depend at all on one of the two inputs. They simply copy or invert the value of one of the inputs. In Table 3.2, these are the $\mathcal{C}_1$ functions. The second class of canalyzing functions has three times a 1 or three times a 0 in its output (the $\mathcal{C}_2$ functions). For each of the two inputs there exists one value that fixes the output irrespective of the other input. In fact, constant functions can also be considered as canalyzing functions, because the output is fixed for any value of the inputs.

Table 3.2 The 16 update functions for nodes with 2 inputs. The first column lists the 4 possible states of the two inputs, the other columns represent one update function each, falling into four classes.

In	$\mathcal{F}$		$\mathcal{C}_1$				$\mathcal{C}_2$								$\mathcal{R}$	
00	1	0	0	1	0	1	1	0	0	0	0	1	1	1	1	0
01	1	0	0	1	1	0	0	1	0	0	1	0	1	1	0	1
10	1	0	1	0	0	1	0	0	1	0	1	1	0	1	0	1
11	1	0	1	0	1	0	0	0	0	1	1	1	1	0	1	0

Each node in the network is assigned an update function by randomly choosing the function from all possible functions with K inputs according to some probability distribution. The simplest probability distribution is a constant one. For $K = 2$ networks, each function is then chosen with probability $1/16$. In the previous section, we have introduced the concept of an ensemble of networks. If we are only interested in topology, an ensemble is defined by the values of N and K. When we want to study network dynamics, we have to assign update functions to each network, and the ensemble needs to be specified by also indicating which probability distribution of the update functions shall be used. If all 4 update functions are allowed, there are 36 different networks in the ensemble shown in Fig. 3.1. For topologies (a) and (b), there are 10 different possibilities to assign update functions, for topology (c) there are 16 different possibilities. The determination of the statistical weight of each of the 36 networks for the case that every update function is chosen with the same probability is left to the reader.

In the following we list several frequently used probability distributions for the update functions. Throughout this article, we will refer to these different "update rules".

1. Biased functions: A function with n times the output value 1 and $M - n$ times the output value 0 is assigned a probability $p^n(1-p)^{M-n}$. Then the two frozen functions in Table 3.2 have the probabilities p^4 and $(1-p)^4$, each of the $\mathcal{C}_1$ functions and of the reversible functions has the probability $p^2(1-p)^2$, and the $\mathcal{C}_2$ functions have the probabilities $p(1-p)^3$ and $p^3(1-p)$. For the special case $p = 1/2$, all functions have the same probability $1/16$.

2. Weighted classes: All functions in the same class are assigned the same probability. $K = 1$ networks are most interesting if the two reversible functions occur with probability $1/2$ each, and the two constant functions with probability 0. In general $K = 1$ networks, we denote the weight of the constant functions with δ. An ensemble of $K = 2$ networks is specified by the four parameters α, β, γ, and δ for the weight of $\mathcal{C}_1$, reversible, $\mathcal{C}_2$ and frozen functions. The sum of the four weights must be 1, i.e., $1 = \alpha + \beta + \gamma + \delta$.

3. Only canalyzing functions are chosen, often including the constant functions. This is motivated by the finding that gene regulation networks appear to have many canalyzing functions and by considerations that canalyzing functions are biologically meaningful [4, 5]. Several authors create canalyzing networks using three parameters [6]. One input of the node is chosen at random to be a canalyzing input. The first parameter, η, is the probability that this input is canalyzing if its value is 1. The second parameter, r, is the probability that the output is 1 if the input is on its canalyzing value. The third parameter, p, assigns update functions for the $K - 1$ other inputs according to rule 1 (biased functions), for the case that the canalyzing input is not on its canalyzing value. (This notation is not uniform throughout literature. For instance, in [6], the second and third parameter are named ρ_1 and ρ_2.)

4. Only threshold functions are chosen, i.e. the update rule is

$$\sigma_i(t+1) = \begin{cases} 1 \text{ if } \sum_{j=1}^{N} \left(c_{ij}(2\sigma_j - 1) + h \right) \geq 0 \\ 0 \text{ else} \end{cases} \tag{3.2}$$

   The couplings c_{ij} are zero if node i receives no input from node j, and they are ± 1 with equal probability if node j is an input to node i. Negative couplings are *inhibitory*, positive couplings are *excitatory*. The parameter h is the *threshold*. Threshold networks are inspired by neural networks, but they are also used in some models for gene regulation networks [7–9].

5. All nodes are assigned the same function. The network is then a cellular automaton with random wiring.

3.2.3
Dynamics

Throughout this paper, we only consider the case of parallel update. All nodes are updated at the same time according to the state of their inputs and to their update function. Starting from some initial state, the network performs a trajectory in state space and eventually arrives on an *attractor*, where the same sequence of states is periodically repeated. Since the update rule is deterministic, the same state must always be followed by the same next state. If we represent the network states by points in the 2^N-dimensional state space, each of these points has exactly one "output", which is the successor state. We thus obtain a graph in state space.

The *size* or *length of an attractor* is the number of different states on the attractor. The *basin of attraction* of an attractor is the set of all states that eventually end up on this attractor, including the attractor states themselves. The size of the basin of attraction is the number of states belonging to it. The graph of states in state space consists of unconnected components, each of them being a basin of attraction and containing an attractor, which is a loop in state space. The *transient* states are those that do not lie on an attractor. They are on *trees* leading to the attractors.

Let us illustrate these concepts by studying the small $K = 1$ network shown in Fig 3.2, which consists of 4 nodes:

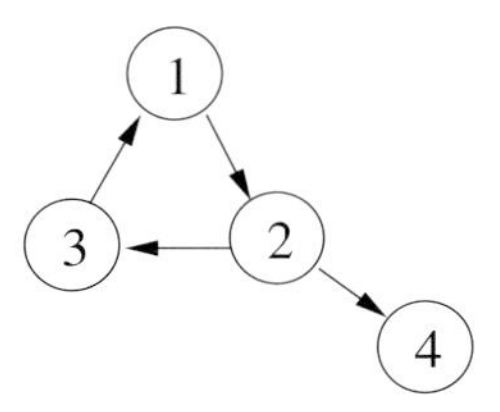

Fig. 3.2 A small network with $K = 1$ input per node.

If we assign to the nodes 1, 2, 3, 4 the functions invert, invert, copy, copy, an initial state 1111 evolves in the following way:

$$1111 \rightarrow 0011 \rightarrow 0100 \rightarrow 1111$$

This is an attractor of period 3. If we interpret the bit sequence characterizing the state of the network as a number in binary notation, the sequence of states can also be written as

$$15 \rightarrow 3 \rightarrow 4 \rightarrow 15$$

The entire state space is shown in Fig. 3.3.

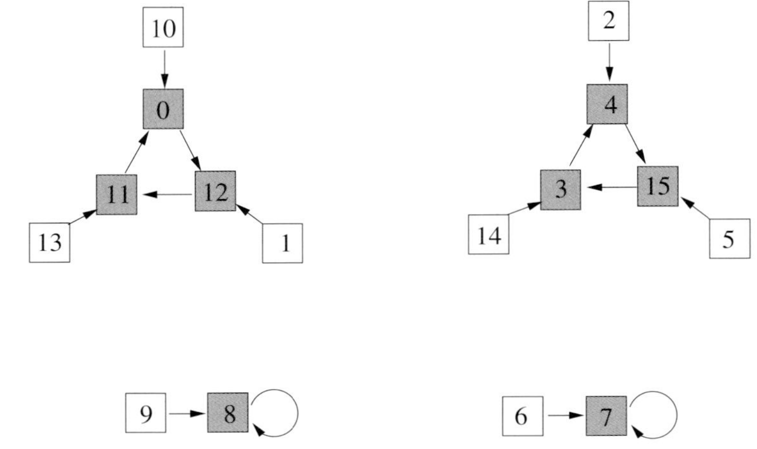

Fig. 3.3 The state space of the network shown in Fig. 3.2, if the functions copy, copy, invert, invert are assigned to the four nodes. The numbers in the squares represent states, and arrows indicate the successor of each state. States on attractors are shaded.

There are 4 attractors, two of which are fixed points (i.e., attractors of length 1). The sizes of the basins of attraction of the 4 attractors are 6,6,2,2. If the function of node 1 is a constant function, fixing the value of the node at 1, the state of this node fixes the rest of the network, and there is only one attractor, which is a fixed point. Its basin of attraction is of size 16. If the functions of the other nodes remain unchanged, the state space then looks as shown in Fig. 3.4.

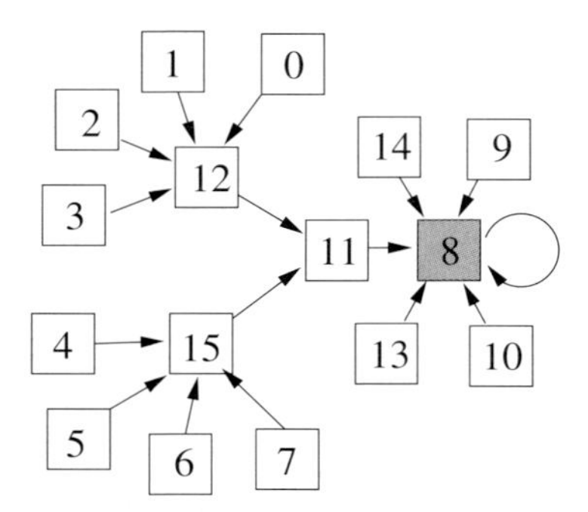

Fig. 3.4 The state space of the network shown in Fig. 3.2, if the functions 1, copy, invert, invert are assigned to the four nodes.

Before we continue, we have to make the definition of attractor more precise: as the name says, an attractor "attracts" states to itself. A periodic sequence of states (which we also call *cycle*) is an attractor if there are states outside the attractor that lead to it. However, some networks contain cycles that cannot be reached from any state that is not part of it. For instance, if we removed node 4 from the network shown in Fig. 3.2, the state space would only contain the cycles shown in Fig. 3.3, and not the 8 states leading to the cycles. In the following, we will use the word "cycle" whenever we cannot be confident that the cycle is an attractor.

3.2.4 Applications

Let us now make use of the definitions and concepts introduced in this chapter in order to derive some results concerning cycles in state space. First, we prove that in an ensemble of networks with update rule 1 (biased functions) or rule 2 (weighted classes), there is on an average exactly one fixed point per network. A fixed point is a cycle of length 1. The proof is slightly different for rule 1 and rule 2. Let us first choose rule 2. We make use of the property that for every update function the inverted function has the same probability. The inverted function has all 1s in the output replaced with 0s, and vice versa. Let us choose a network state, and let us determine for which fraction of networks in the ensemble this state is a fixed point. We choose a network at random, prepare it in the chosen state, and perform one update step. The probability that node 1 remains in the same state after the update, is 1/2, because a network with the inverted function at node 1 occurs equally often. The same holds for all other nodes, so that the chosen state is a fixed point of a given network with probability 2^{-N}. This means that each of the 2^N states is a fixed point in the proportion 2^{-N} of all networks, and therefore the mean number of fixed points per network is 1. We will see later that fixed points may be highly clustered: a small proportion of all networks may have many fixed points, while the majority of networks have no fixed point.

Next, we consider rule 1. We make now use of the property that for every update function a function with any permutation of the input states has the same probability. This means that networks in which state A leads to state B after one update, and networks in which another state C leads to state B after one update, occur equally often in the ensemble. Let us choose a network state with n 1s and $N - n$ 0s. The average number of states in a network leading to this state after one update is $2^N p^n (1-p)^{N-n}$. Now, every state leads equally often to this state, and therefore this state is a fixed point in the proportion $p^n(1-p)^{N-n}$ of all networks. Summation over all states gives the mean number of fixed points per network, which is 1.

Finally, we derive a general expression for the mean number of cycles of length L in networks with $K = 2$ inputs per node. The generalization to other values of K is straightforward. Let $\langle C_L \rangle_N$ denote the mean number of cycles in state space of length L, averaged over the ensemble of networks of size N. On a cycle of length L, the state of each node goes through a sequence of 1s and 0s of period L. Let us number the 2^L possible sequences of period L of the state of a node by the index j, ranging from 0 to $m = 2^L - 1$. Let n_j denote the number of nodes that have the sequence j on a cycle of length L, and $(P_L)^j_{l,k}$ the probability that a node that has the input sequences l and k generates the output sequence j. This probability depends on the probability distribution of update functions. Then

$$\langle C_L \rangle_N = \frac{1}{L} \sum_{\{n_j\}} \frac{N!}{n_0! \ldots n_m!} \prod_j \left(\sum_{l,k} \frac{n_l n_k}{N^2} (P_L)^j_{l,k} \right)^{n_j} . \tag{3.3}$$

The factor $1/L$ occurs because any of the L states on the cycle could be the starting point. The sum is over all possibilities to choose the values $\{n_j\}$ such that $\sum_j n_j = N$. The factor after the sum is the number of different ways in which the nodes can be divided into groups of the sizes $n_0, n_1, n_2, \ldots, n_m$. The product is the probability that each node with a sequence j is connected to nodes with the sequences l and k and has an update function that yields the output sequence j for the input sequences l and k. This formula was first given in the beautiful paper by Samuelsson and Troein [10].

We conclude this chapter with a picture of the state space of a network consisting of 10 nodes.

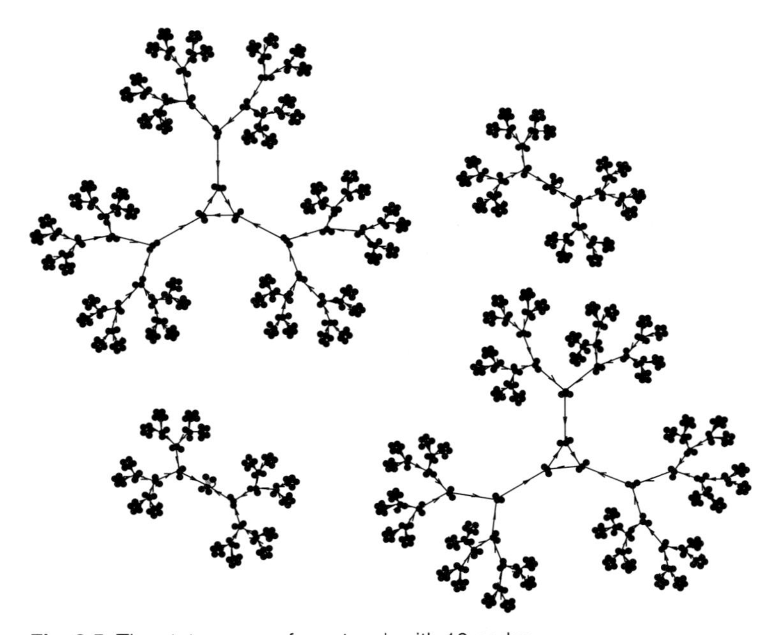

Fig. 3.5 The state space of a network with 10 nodes.

3.2.5
Problems

1. Show that the fraction $3/32$ of all networks in the ensemble with $N = 4$ and $K = 1$ have the topology shown in Fig. 3.2.

2. Show that the fraction $3^3/2^{10}$ of all networks with the topology shown in Fig. 3.2 have the state space topology shown in Fig. 3.3, if the distribution of update functions is given by rule 1 with $p = 1/4$.

3. Which functions in Table 3.2 correspond to the threshold functions in networks with $K = 2$, if we set $h = 0$?

4. Consider again $K = 2$ networks, and choose the update rules 3 (canalyzing functions), which are characterized by the parameters η, r, and p. Express the weight of each function in Table 3.2 in terms of η, r, and p.

5. Using Eq. (3.3), show that in an ensemble of networks with update rule 1 or 2, there is on an average exactly one fixed point per network.

3.3 Annealed Approximation and Phase Diagrams

The *annealed approximation*, which is due to Derrida and Pomeau [11], is a useful tool to calculate certain network properties. It is a mean-field theory, which neglects possible correlations between nodes. The first assumption of the annealed approximation is that the network is infinitely large. This means that fluctuations of global quantities are negligible. The second assumption of the annealed approximation is that the inputs of each node can be assigned at every time step anew. The following quantities can be evaluated by the annealed approximation:

1. The time evolution of the proportion of 1s and 0s.
2. The time evolution of the Hamming distance between the states of two identical networks.
3. The statistics of small perturbations.

We will discuss these in the following in the order given in this list. One of the main results of these calculations will be the phase diagram, which indicates for which parameter values the networks are frozen, critical or chaotic.

3.3.1 The Time Evolution of the Proportion of 1s and 0s

Let b_t denote the number of nodes in state 1, divided by N. The proportion of nodes in state 0 is then $1 - b_t$. We want to calculate b_{t+1} as function of b_t within the annealed approximation. Since the K inputs of each node are newly assigned at each time step, the probability that m inputs of a node are in state 1 and the other inputs in state 0 is $b_t^m(1 - b_t)^{K-m}$. Since we consider an infinitely large network, this probability is identical to the proportion of nodes that have m inputs in state 1.

Let p_m be the probability that the output value of a node with m inputs in state 1 is 1. Then we have

$$b_{t+1} = \sum_{m=0}^{K} \binom{K}{m} p_m b_t^m (1 - b_t)^{K-m} \,. \tag{3.4}$$

If p_m is independent of m, the right-hand side is identical to p_m, and b_t reaches after one time step its stationary value, which is the fixed point of

Eq. (3.4). Among the above-listed update rules, this happens for rules 1 (biased functions) and 2 (weighted classes) and 4 (threshold functions). For rule 1, we have $p_m = 1/2$, since the output values 0 and 1 occur with equal probability within each class of update functions. For rule 2, we have $p_m = p$ by definition. For rule 4, the value of p_m is independent of m because the value of c_{ij} is 1 and -1 with equal probability, making each term $c_{ij}(2\sigma_j - 1)$ to $+1$ and -1 with equal probability. Therefore p_m is identical to the probability that the sum of K random numbers, each of which is $+1$ or -1 with probability $1/2$, is at least as large as $-h$,

$$p_m = \left(\frac{1}{2}\right)^K \sum_{l \geq (K-h)/2} \binom{K}{l} .$$

Here, l is the number of $+1$s, and $K - l$ the number of -1s.

For rule 3 (canalyzing functions) we get [6]

$$\begin{aligned} b_{t+1} &= b_t \eta r + (1 - b_t)(1 - \eta) r \\ &\quad + b_t (1 - \eta) p + (1 - b_t) \eta p \\ &= r + \eta(p - r) + b_t (p - r)(1 - 2\eta) . \end{aligned} \tag{3.5}$$

The first two terms are the probability that the canalyzing input is on its canalyzing value, and that the output is then 1. The second two terms are the probability that the canalyzing input is not on its canalyzing value, and that the output is then 1. This is a one-dimensional map. The only fixed point of this map is

$$b^* = \frac{r + \eta(p - r)}{1 - (p - r)(1 - 2\eta)} .$$

Since the absolute value of the slope of this map is smaller than 1 everywhere, every iteration (3.5) will bring the value of b_t closer to this fixed point.

There exist also update rules where the fixed points are unstable and where periodic oscillations or chaos occur. This occurs particularly easily when all nodes are assigned the same function (rule 5). For instance, if all nodes are assigned the last one of the canalyzing functions occurring in the table of update functions 3.2, we have the map

$$b_{t+1} = 1 - b_t^2 . \tag{3.6}$$

The fixed point

$$b^* = \frac{-1 + \sqrt{5}}{2}$$

is unstable, since the slope of the map is $(1-\sqrt{5})$ at this fixed point, i.e., it has an absolute value larger than 1. The iteration (3.6) moves b_t away from this fixed point, and eventually the network oscillates between all nodes being 1 and all nodes being 0.

A map that allows for oscillations with larger period and for chaos is obtained for the update rule that the output is 1 only if all inputs are equal. This map is defined for general values of K and is given by

$$b_{t+1} = b_t^K + (1-b_t)^K \,. \tag{3.7}$$

Let us consider K as a continuous parameter. When it is increased, starting at 1, the map first has a stable fixed point and then shows a period-doubling cascade and the Feigenbaum route to chaos shown in Fig. 3.6 [12].

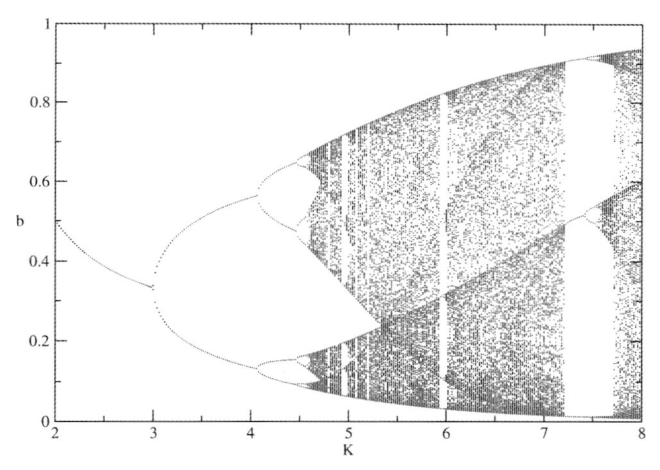

Fig. 3.6 The values of b_t that still occur after the transient time for the map (3.7), as function of K.

All these results for b_t were derived within the annealed approximation, but they are generally believed to apply also to the original networks with fixed connectivity patterns, if the thermodynamic limit is taken. If this is correct, the following three statements are also correct:

- All (apart from a vanishing proportion of) initial states with a given value of b_0 undergo the same trajectory b_t with time.
- This trajectory is the same for all networks (apart from a vanishing proportion).
- When time is so large that the dynamics have reached an attractor, the map $b_{t+1}(b_t)$ is the same as in the initial stage for those values of b that can occur on the attractors.

These assumptions appear plausible, since the paths through which a node can affect its own input nodes are infinitely long in a randomly wired, infinitely large network. Therefore we do not expect correlations between the update function assigned to a node and the states of its input nodes. Neither do we expect a correlation between the function $b_{t+1}(b_t)$ and the question of whether a state is on an attractor.

3.3.2
The Time Evolution of the Hamming Distance

With the help of the Hamming distance, one can distinguish between a frozen and a chaotic phase for RBNs. We make an identical copy of each network in the ensemble, and we prepare the two copies of a network in different initial states. The Hamming distance between the two networks is defined as the number of nodes that are in a different state. For the following, it is more convenient to use the *normalized Hamming distance*, which is the Hamming distance divided by N, i.e., the proportion of nodes that are in a different state,

$$h_t = \frac{1}{N} \sum_{i=1}^{N} \left(\sigma_i^{(1)} - \sigma_i^{(2)} \right)^2 . \tag{3.8}$$

If h_t is very small, the probability that more than one input of a node differ in the two copies, can be neglected, and the change of h_t during one time step is given by

$$h_{t+1} = \lambda h_t , \tag{3.9}$$

where λ is called the *sensitivity* [13]. It is K times the probability that the output of a node changes when one of its inputs changes.

For the first four update rules listed in Section 3.2.2 the value of λ is

$$\lambda = 2Kp(1-p) \qquad \text{(biased functions)}$$

$$\lambda = 1 - \delta \qquad \text{(weighted classes, } K = 1\text{)}$$

$$\lambda = \alpha + 2\beta + \gamma = 1 + \beta - \delta \qquad \text{(weighted classes, } K = 2\text{)}$$

$$\lambda = r(1-p) + (1-r)p + (K-1)(\eta(1-b_t) + (1-\eta)b_t)2p(1-p) \qquad \text{(canalyzing functions)}$$

$$\lambda = K \left(\frac{1}{2}\right)^{K-1} \binom{K-1}{l} \qquad \text{(threshold functions)} \tag{3.10}$$

with l in the last line being the largest integer smaller than or equal to $(K-h)/2$. For rule 3 (canalyzing functions), the first two terms are the probability that the output changes when the canalyzing input is in a different state in the two network copies; the last term is the probability that the output changes when one of the other inputs is in a different state in the two copies, multiplied by the number of noncanalyzing inputs. This is the only one out of the 4 rules where the value of λ depends on b_t and therefore on time.

The networks are in different phases for $\lambda < 1$ and $\lambda > 1$, with the critical line at $\lambda = 1$ separating the two phases. In the following, we derive the properties of the networks in the two phases as far as possible within the annealed approximation.

If $\lambda < 1$, the normalized Hamming distance decreases to 0. If the states of the two copies differ initially in a small proportion of all nodes, they become identical for all nodes, apart from possibly a limited number of nodes, which together make a contribution 0 to the normalized Hamming distance. $\lambda < 1$ means also that if the two copies are initially in identical states and the state of one node in one copy is changed, this change propagates on an average to less than one other node. When the two copies differ initially in a larger proportion of their nodes, we can argue that their states also become identical after some time: we produce a large number Q of copies of the same network and prepare their initial states such that copy number q and copy number $q + 1$ (for all $q = 1, ..., Q$) differ only in a small proportion of their nodes. Then the states of copy number q and copy number $q + 1$ will become identical after some time, and therefore the states of all Q copies become identical (again apart from possibly a limited number of nodes). The final state at which all copies arrive must be a state where all nodes (apart from possibly a limited number) become frozen at a fixed value. If the final state was an attractor where a nonvanishing proportion of nodes go through a sequence of states with a period larger than 1, different network copies could be in different phases of the attractor, and the normalized Hamming distance could not become zero. Ensembles with $\lambda < 1$ are said to be in the *frozen* phase.

All these considerations did not take into account that λ itself may not be constant. For those update rules where b_t assumes its fixed point value after the first time step, one can apply the reasoning of the previous paragraph starting at time step 2. For rule 3, the value b_t approaches its fixed point more slowly, and therefore the value of λ changes over a longer time period. It is therefore possible that the Hamming distance shows initially another trend as during later times. Once b_t has reached its fixed point value, λ has become constant, and if then $\lambda < 1$, the normalized Hamming distance will decrease to zero. In order to decide whether an ensemble is in the frozen phase, one must therefore evaluate λ in the stationary state.

For ensembles that have no stable stationary value of b_t, the above considerations do not apply directly, since b_t cycles through different values, and so does λ. Furthermore, the two copies may be in different phases of the cycle and will then never have a small normalized Hamming distance. For ensembles with a finite oscillation period T, one should evaluate the product of all values of λ during one period. If this product is smaller than 1, a small normalized Hamming distance created in a copy of a network with a stationary oscillation, will decrease after one period. Using a similar reasoning as before, we conclude that then the normalized Hamming distance between any two copies of the network will decrease to zero if they have initially the same value of b_t. This means that all nodes (apart from possibly a limited number) go through a sequence of states that has the same period as b_t.

If $\lambda > 1$ when b_t has reached its stationary value, the normalized Hamming distance increases from then on with time and has a nonzero stationary value. A change in one node propagates on an average to more than one other node. If there is a fixed point or a short attractor, it is unstable under many possible perturbations. There is therefore no reason why all attractors should be short. In fact, attractors can be very long, and the ensemble is in a phase that

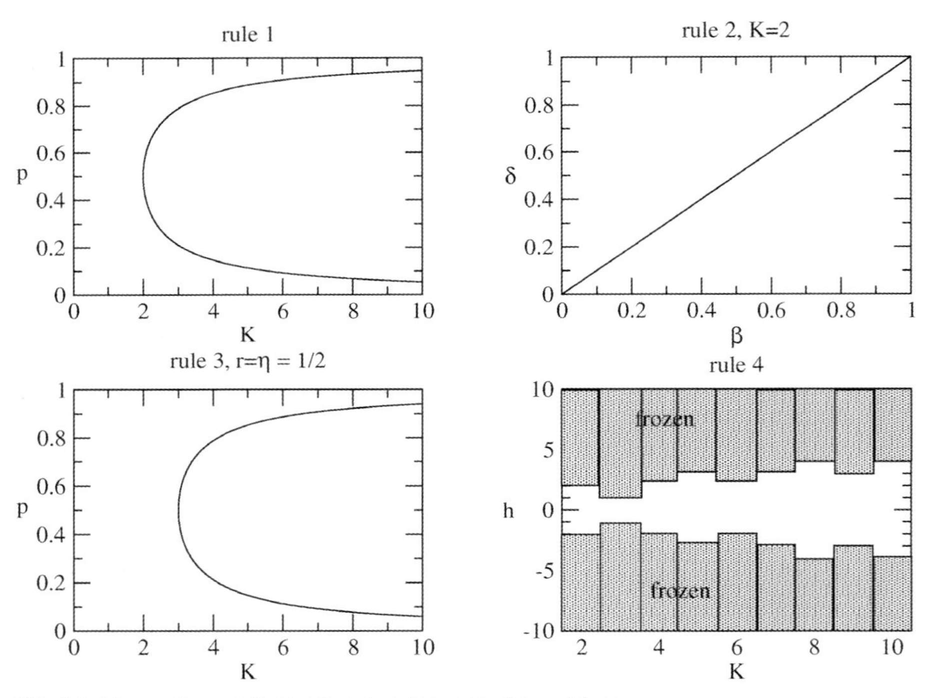

Fig. 3.7 Phase diagram for the first 4 update rules (biased functions, weighted classes, canalyzing functions, threshold functions). Where there are more than 2 parameters, the remaining parameters were fixed at the values given in the respective graph titles. For threshold functions, the frozen phase is shaded. For $K = 2$, the model is critical between $h = -2$ and 2, for larger K, there exists no critical value, but the model is chaotic whenever it is not frozen.

is usually called *chaotic*, even though this is no real chaos because state space is finite and every trajectory becomes eventually periodic. When b_t does not become stationary but periodic, we consider again the product of all values of λ during one period. If this product is larger than 1, a small normalized Hamming distance between two copies with the same value of b_t will eventually become larger. This means that attractors can be very long and need not have the period of b_t.

For $\lambda = 1$, the ensemble is at the boundary between the two phases: it is critical. A change in one node propagates on an average to one other node. The critical line can be obtained from Eqs. (3.10), leading to the phase diagram shown in Fig. 3.7.

All our results are based on a calculation for small h_t. When h_t is not infinitesimally small, Eq. (3.9) has the more general form

$$h_{t+1} = \lambda h_t + \nu h_t^2 + \dots, \tag{3.11}$$

with the highest power of h_t being K, but we do not make here the effort to calculate the coefficient ν or that of a higher-order term. We use this result only to obtain a relation between the stationary value of h_t and the distance from the critical line: In the chaotic phase, but close to the critical line (where λ is only slightly larger than 1), the stationary value of h_t obtained from Eq. (3.11) is

$$h^* = (\lambda - 1)/\nu \,. \tag{3.12}$$

It increases linearly with the distance from the critical line, as long as this distance is small.

3.3.3
The Statistics of Small Perturbations in Critical Networks

Now let us have a closer look at the propagation of a perturbation that begins at one node in a critical network. Let us consider again two identical networks, and let them be initially in the same state. Then let us flip the state of one node in the first network. One time step later, the nodes that receive input from this node differ in the two systems each with probability $\lambda/K = 1/K$ (since $\lambda = 1$ in a critical network). On an average, this is one node. Since the perturbation propagates to each node in the network with probability $(K/N) * (\lambda/K) = 1/N$, the probability distribution is a Poisson distribution with mean value 1. We keep track of all nodes to which the perturbation propagates, until no new node becomes affected by it. We denote the total number of nodes affected by the perturbation by s. The size distribution of perturbations is a power law

$$n(s) \sim s^{-3/2} \tag{3.13}$$

for values of s that are so small that the finite system size is not yet felt, but large enough to see the power law. There are many ways to derive this power law. The annealed approximation consists in assuming that loops can be neglected, so that the perturbation propagates at every step through new bonds and to new nodes. In this case, there is no difference (from the point of view of the propagating perturbation) between a network where the connections are fixed and a network where the connections are rewired at every time step.

We begin our calculation with one "active" node at time 0, $n_a(t = 0) = 1$, which is the node that is perturbed. At each "time step" (which is different from real time!), we choose one active node and ask to how many nodes the perturbation propagates from this node in one step. These become active nodes, and the chosen node is now "inactive". We therefore have a stochastic process

$$n_a(t+1) = n_a(t) - 1 + \xi$$

for the number of "active" nodes, with ξ being a random number with a Poisson distribution with mean value 1. The stochastic process is finished at time $t = s$ when $n_a(s) = 0$. s is the total number of nodes affected by the perturbation.

Now we define $P_0(y,t)$ as the probability that the stochastic process has arrived at $y = 0$ before or at time t, if it has started at $n_a = y$ at time $t = 0$. During the first step, y changes by $\Delta y = \xi - 1$. If we denote the probability distribution of Δy with $P(\Delta y)$, we obtain

$$\begin{aligned} P_0(y,t) &= \int d(\Delta y) P(\Delta y) P_0(y + \Delta y, t-1) \\ &\simeq \int d(\Delta y) P(\Delta y) \left[P_0(y,t) + \Delta y \frac{\partial P_0}{\partial y} + \frac{1}{2}(\Delta y)^2 \frac{\partial^2 P_0}{\partial y^2} - \frac{\partial P_0}{\partial t} \right] . \end{aligned}$$

The first term on the right-hand side cancels the left-hand side. The second term on the right-hand side is the mean value of Δy, which is zero, times $\partial_y P_0$. Were are therefore left with the last two terms, which give after integration

$$\frac{\partial P_0}{\partial t} = \frac{1}{2} \frac{\partial^2 P_0}{\partial y^2} . \tag{3.14}$$

This is a diffusion equation, and we have to apply the initial and boundary conditions

$$\begin{aligned} P_0(0,t) &= 1 \\ P_0(y,0) &= 0 \\ P_0(y,\infty) &= 1 \end{aligned} \tag{3.15}$$

Expanding P_0 in terms of eigenfunctions of the operator $\partial/\partial t$ gives the general solution

$$P_0(y,t) = a + by + \int d\omega\, e^{-\omega^2 t/4}\left(c_\omega \sin(\omega y) + d_\omega \cos(\omega y)\right).$$

The initial and boundary conditions fix the constants to $a = 1$ and $d_\omega = 0$ and $c_\omega = -2/\pi\omega$. We therefore have

$$P_0(y,t) = 1 - \frac{2}{\pi}\int d\omega \frac{\sin \omega y}{\omega}\, e^{-\omega^2 t/4}, \tag{3.16}$$

which becomes for $y = 1$

$$\begin{aligned} P_0(1,t) &= 1 - \frac{2}{\pi}\int d\omega\, \frac{\sin \omega}{\omega}\, e^{-\omega^2 t/4} \\ &\to 1 - \mathcal{O}(t^{-1/2}) \end{aligned} \tag{3.17}$$

for large t. The size distribution of perturbations is obtained by taking the derivative with respect to t, leading to Eq. (3.13).

Readers familiar with percolation theory will notice that the spreading of a perturbation in a critical RBN is closely related to critical percolation on a Bethe lattice. Only the probability distribution of the stochastic variable ξ is different in this case. Since the result depends only on the existence of the second moment of y, it is not surprising that the size distribution of critical percolation clusters on the Bethe lattice follows the same power law.

3.3.4 Problems

1. Explain why there is a finite critical region for $K = 2$ and no critical value of λ at all for $K > 2$ for update rule 4 (see Fig. 3.7).

2. For each of the 16 update functions for $K = 2$, consider an ensemble where all nodes are assigned this function. Find the function $b_{t+1}(b_t)$. Find all fixed points of b and determine if they are stable. If b_t becomes constant for large times, determine whether the ensemble is frozen, critical or chaotic. If b_t oscillates for large times, determine whether the normalized Hamming distance between two identical networks that start with the same value of b_0, goes to zero for large times. Interpret the result.

3. If in a frozen network only a limited number of nodes may not be frozen for large times, and if in a chaotic network a nonvanishing proportion of nodes remain nonfrozen, what do you expect in a critical network?

3.4
Networks with $K = 1$

Many properties of networks with $K = 1$ inputs per node can be derived analytically. Nevertheless, these networks are nontrivial and share many features with networks with larger values of K. Therefore it is very instructive to have a closer look at $K = 1$ networks. In this review, we will not reproduce mathematically exact results that require long calculations, as is for instance done in [14, 15]. Instead, we will present phenomenological arguments that reproduce correctly the main features of these networks and that help to understand how these features result from the network structure and update rules. We begin by studying the topology of $K = 1$ networks. Then, we will investigate the dynamics on these networks in the frozen phase and at the critical point. Finally, we will show that the topology of $K = 1$ networks can be mapped on the state space of $K = N$ networks, which allows us to derive properties of the attractors of $K = N$ networks, which are chaotic.

3.4.1
Topology of $K = 1$ Networks

If each node has one input, the network consists of different components, each of which has one loop and trees rooted in this loop, as shown in Fig. 3.8. $K = 1$ networks have the same structure as the state space pictures of other random Boolean networks, like the ones shown in Figs. 3.3 and 3.4, only the arrows are inverted.

Let us first calculate the size distribution of loops. We consider the ensemble of all networks of size N. In each network of the ensemble, each node chooses its input at random from all other nodes. The probability that a given node is

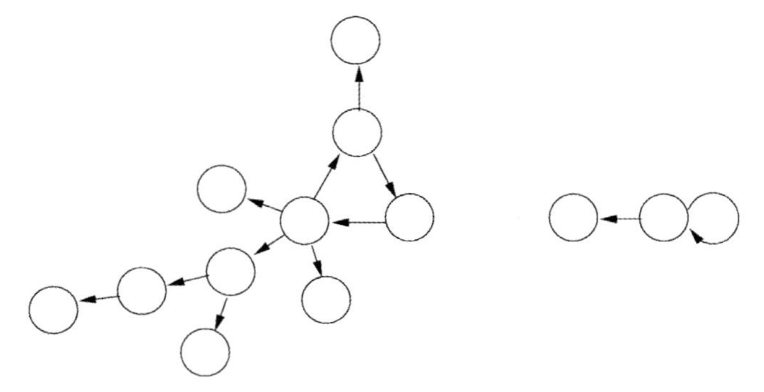

Fig. 3.8 Example of a network with one input per node. It has two components, the larger component has a loop of size 3 and two trees rooted in it (one of size 1 and one of size 6), and the smaller component has a loop of size 1 and one tree of size 1.

sitting on a loop of size l is therefore

$$P(l) = \left(1-\frac{1}{N}\right)\left(1-\frac{2}{N}\right)\cdots\left(1-\frac{l-1}{N}\right)\frac{1}{N}$$
$$\simeq \frac{e^{-1/N}e^{-2/N}\ldots e^{-(l-1)/N}}{N}$$
$$= \frac{e^{-l(l-1)/2N}}{N} \simeq \frac{e^{-l^2/2N}}{N}\,. \tag{3.18}$$

The first factor is the probability that the input to the first node is not this node. The second factor is the probability that the input to the second node is not the first or second node, etc. The last factor is the probability that the input of the lth node is the first node. The approximation in the second step becomes exact in the thermodynamic limit $N \to \infty$ for values of l that satisfy $\lim_{N\to\infty} l/N = 0$. The approximation in the last step can be made if l is large.

The probability that a given node is sitting on any loop is therefore proportional to

$$\int_1^\infty P(l)dl \simeq N^{-1/2}\int_0^\infty e^{-x^2/2}dx \propto N^{-1/2}\,.$$

This means that the total number of nodes sitting on loops is proportional to $\sqrt{N}$.

The mean number of nodes sitting on loops of size l in a network is

$$NP(l) \simeq e^{-l^2/2N}\,.$$

The cutoff in loop size is proportional to $\sqrt{N}$. For $l \ll \sqrt{N}$, the mean number of nodes in loops of size l is 1. This result can also be obtained by a simple argument: The probability that a given node is sitting on a loop of size l is in the limit $N \to \infty$ simply $1/N$, since the node almost certainly does not choose itself as input or as input of its input etc, but in the lth step the first node must be chosen as input, which happens with probability $1/N$.

The mean number of loops of size l in a network is

$$\frac{NP(l)}{l} \simeq e^{-l^2/2N}/l\,.$$

For $l \ll \sqrt{N}$, this is simply $1/l$. Since loops are formed independently from each other in the limit $N \to \infty$, the probability distribution of the number of loops of size l is a Poisson distribution with mean value $1/l$.

The mean number of loops per network is

$$\sum_l NP(l)/l \simeq \int_{N^{-1/2}}^{\infty} \frac{e^{-x^2/2}}{x} dx \simeq \frac{1}{2} \ln N$$

for large N. This is identical to the mean number of components.

Next, let us consider the trees rooted in the loops. There are of the order of N nodes, which sit in $\propto \sqrt{N}$ trees, each of which is rooted in a relevant node. This means that the average tree size is proportional to $\sqrt{N}$. The construction of a tree can be described formally exactly in the same way as we described the propagation of a perturbation in a critical network in the last chapter: we begin with a node sitting in a loop. The nodes that are not sitting in loops receive their input with equal probability from any node in the network. Our node is therefore chosen with probability $1/N$ by every node outside the loops as an input, and the probability distribution of the number of outputs into the tree is a Poisson distribution with mean value 1 (neglecting terms of the order $N^{-1/2}$). In the same way, we find that the number of outputs of each of the newly found tree nodes is again a Poisson distribution with mean value 1. We iterate this process until we have identified all nodes that are part of this tree.

The size distribution of trees is $\sim s^{-3/2}$. The cutoff must be $s_{max} \sim N$ in order to be consistent with what we know about the mean tree size and the total number of nodes in trees: The mean tree size is

$$\bar{s} \sim \int_1^{s_{max}} s s^{-3/2} ds \sim s_{max}^{1/2} \sim \sqrt{N}\,.$$

The total number of nodes in trees is proportional to

$$\sqrt{N} \int_1^{s_{max}} s s^{-3/2} ds \sim N\,.$$

3.4.2
Dynamics on $K = 1$ Networks

Knowing the topology of $K = 1$ networks, allows us to calculate their dynamical properties. After a transient time, the state of the nodes on the trees will be independent of their initial state. If a node on a tree does not have a constant function, its state is determined by the state of its input node at the previous time step. All nodes that are downstream of a node with a constant function will become frozen. If there is no constant function in the loop and the path from the loop to a node, the dynamics of this node is slaved to the dynamics of the loop.

If the weight of constant functions, δ, is nonzero, the probability that a loop of size l does not contain a frozen function is $(1-\delta)^l$, which goes to zero when

l is much larger than $1/\delta$. Therefore only loops smaller than a cutoff size can have nontrivial dynamics.

The number and length of the attractors of the network are determined by the nonfrozen loops only. Once the cycles that exist on each of the nonfrozen loops are determined, the attractors of the entire networks can be found from combinatorial arguments.

3.4.2.1 Cycles on Loops

Let us therefore focus on a loop that has no constant function. If the number of "invert" functions is odd, we call the loop an odd loop. Otherwise it is an even loop. Replacing two "invert" functions with copy functions and replacing the states $\sigma_i(t)$ of the two nodes controlled by these functions and of all nodes in between with $1 - \sigma_i(t)$, is a bijective mapping from one loop to another. In particular, the number and length of cycles on the loop is not changed. All odd loops can thus be mapped on loops with only one "invert" function, and all even loops can be mapped on loops with only "copy" functions.

We first consider even loops with only "copy" functions. These loops have two fixed points, where all nodes are in the same state. If l is a prime number, all other states belong to cycles of period l. Any initial state occurs again after l time steps. Therefore the number of cycles on an even loop is

$$\frac{2^l - 2}{l} + 2 \tag{3.19}$$

if l is a prime number. The numerator counts the number of states that are not fixed points. The first term is therefore the number of cycles of length l. Adding the two fixed points gives the total number of cycles. If l is not a prime number, there exist cycles with all periods that are a divisor of l.

Next, let us consider odd loops with one "invert" function. After $2l$ time steps, the loop is in its original state. If l is a prime number, there is only one cycle that has a shorter period. It is a cycle with period 2, where at each site 0s and 1s alternate. The total number of cycles on an odd loop with a prime number l is therefore

$$\frac{2^l - 2}{2l} + 1\,. \tag{3.20}$$

If l is not a prime number, there are also cycles with a period that is twice a divisor of l.

3.4.2.2 $K = 1$ Networks in the Frozen Phase

For networks with $K = 1$ input per node, the parameter λ is

$$\lambda = 1 - \delta\,. \tag{3.21}$$

Therefore, only networks without constant functions are critical. Networks with $\delta > 0$ are in the frozen phase. The mean number of nonfrozen nodes on nonfrozen loops is given by the sum

$$\sum_l (1-\delta)^l = \frac{1-\delta}{\delta}\,. \tag{3.22}$$

We call these loops the *relevant loops*. We call the nodes on the relevant loops the *relevant nodes*, and we denote their number with N_{rel}. The mean number of relevant loops is given by the sum

$$\sum_l \frac{1}{l}(1-\delta)^l \simeq \ln \delta^{-1}\,, \tag{3.23}$$

with the last step being valid for small δ.

The probability that the activity moves up the tree to the next node is $1-\delta$ at each step. The mean number of nonfrozen nodes on trees is therefore

$$\frac{1-\delta}{\delta}\sum_l (1-\delta)^l = \left(\frac{1-\delta}{\delta}\right)^2\,, \tag{3.24}$$

and the total mean number of nonfrozen nodes is $(1-\delta)/\delta^2$. This is a finite number, which diverges as δ^{-2} when the critical point $\delta = 0$ is approached.

3.4.2.3 **Critical K = 1 Networks**

If the proportion of constant functions δ is zero, the network is critical, and all loops are relevant loops. There are no nodes that are frozen on the same value on all attractors. A loop of size 1 has a state that is constant in time, but in can take two different values. Larger loops have also two fixed points, if they are even. Part of the nodes in a critical $K = 1$ networks are therefore frozen on some attractors or even on all attractors, however, they can be frozen in different states.

The network consists of $\simeq \ln N/2$ loops, each of which has of the order $2^l/l$ cycles of a length of the order l. The size of the largest loop is of the order of $\sqrt{N}$. The number of attractors of the network results from the number of cycles on the loops. It is at least as large as the product of all the cycle numbers of all the loops. If a cycle is not a fixed point, there are several options to choose its phase, and the number of attractors of the network becomes larger than the product of the cycle numbers. An upper bound is the total number of states of all the loops, which is $2^{N_{rel}} \sim e^{a\sqrt{N}}$, and a lower bound is the number of attractors on the largest loop, which is of the order $e^{b\sqrt{N}}/\sqrt{N} > e^{b'\sqrt{N}}$ with $b' < b < a$. From this it follows that the mean number of attractors of critical

$K = 1$ networks increases exponentially with the number of relevant nodes. A complementary result for the number of cycles $\langle C_L \rangle$ of length L, which is valid for fixed L in the limit $N \to \infty$ is obtained by the following quick calculation:

$$\begin{aligned}
\langle C_L \rangle_N &\simeq \sum_{\{n_l\}} \prod_{l \leq l_c} \left(\frac{e^{-1/l} \left(\frac{1}{l}\right)^{n_l}}{n_l!} k_l^{n_l} \right) = \sum_{\{n_l\}} \prod_{l \leq l_c} \left(\frac{e^{-1/l} \left(\frac{k_l}{l}\right)^{n_l}}{n_l!} \right) \\
&\simeq \prod_{l \leq l_c} e^{(k_l - 1)/l} = e^{\int_1^{l_c} (k_l - 1) dl/l} \simeq e^{(\bar{k}_l - 1) \int_1^{l_c} dl/l)} \\
&\sim e^{(H_L - 1) \ln \sqrt{N}} = N^{(H_L - 1)/2} .
\end{aligned} \tag{3.25}$$

Here, n_l is the number of loops of size l, l_c is the cutoff in loop size $\propto \sqrt{N}$, and k_l is the number of states on a loop of size l that belong to a cycle of length L. This is zero for many loops. The average over an l-interval of size L is identical to H_L, which is the number of cycles on an even loop of size L. A more precise derivation of this relation, starting from the $K = 1$ version of Eq. (3.3) and evaluating it by making a saddle-point approximation, can be found in [16], which is inspired by the equivalent calculation for $K = 2$ critical networks in [10].

The length of an attractor of the network is the least common multiple of the cycle lengths of all the loops. A quick estimate gives

$$N^{a \log N}$$

since the length of the larger loops is proportional to $\sqrt{N}$, and this has to be taken to a power which is the number of loops. A more precise calculation [17] gives this expression, multiplied with a factor $N^b / \log N$, which does not modify the leading dependence on N.

3.4.3
Dynamics on K = N Networks

The topology of a $K = 1$ network is identical to the topology of the state space of a $K = N$ network, when all update functions are chosen with the same weight. The reason is that in a $K = N$ network, the state that succeeds a given state can be every state with the same probability. Thus, each state has one successor, which is chosen at random among all states. In the same way, in a $K = 1$ network, each node has one input node, which is chosen at random among all nodes. The state space of a $K = N$ network consists of 2^N nodes, each of which has one successor. We can now take over all results for the topology of $K = 1$ networks and translate them into state space:

The $K = N$ networks have of the order of $\log(2^N) \propto N$ attractors. The largest attractor has a length of the order $\sqrt{2^N} = 2^{N/2}$, and this is proportional to the total number of states on attractors. All other states are transient states. An attractor of length l occurs with probability $1/l$ if $l \ll 2^{N/2}$.

Clearly, $K = N$ networks, where all update functions are chosen with the same weight, are in the chaotic phase. The mean number of nodes to which a perturbation of one node propagates, is $N/2$. At each time step, half the nodes change their state, implying also that the network is not frozen.

3.4.4
Application: Basins of Attraction in Frozen, Critical and Chaotic Networks

The advantage of $K = 1$ networks is that they are analytically tractable and can teach us at the same time about frozen, chaotic and critical behavior. We will discuss in the next chapter to what extent the results apply to networks with other values of K. Based on our insights into $K = 1$ networks, we derive now expressions for the dependence on N of the number and size of the basins of attraction of the different attractors.

Let us first consider networks in the frozen phase. As we have seen, there is at most a limited number of small nonfrozen loops. Their number is independent of system size, and therefore the number of attractors is also independent of the system size. The initial state of the nodes on these nonfrozen loops determines the attractor. The initial states of all other nodes are completely irrelevant at determining the attractor.

The size of the basin of attraction of an attractor is therefore $2^{N-N_{rel}}$, multiplied with the length of the attractor, i.e., it is 2^N, divided by a factor that is independent of N. The proportion of state space belonging to a basin is therefore also independent of N. If we define the *basin entropy* [18] by

$$S = -\sum_a p_a \ln p_a \tag{3.26}$$

with p_a being the fraction of state space occupied by the basin of attraction of attractor a, we obtain

$$S = const$$

for a $K = 1$ network in the frozen phase.

Next, let us consider the chaotic $K = N$ network ensemble. There are on an average $1/l$ attractors of length l, with a cutoff around $2^{N/2}$. The basin size of an attractor of length l is of the order $l2^{N/2}$, which is l times the average tree size. The basin entropy is therefore

$$S \simeq \sum_l \frac{1}{l} \frac{l}{2^{N/2}} \log \frac{l}{2^{N/2}} \simeq \int_{2^{-N/2}}^{1} \log x dx = const\,. \tag{3.27}$$

Finally, we evaluate the basin entropy for a critical $K = 1$ network. There are of the order $e^{a\sqrt{N}}$ attractors with approximately equal basin sizes, and therefore the basin entropy is

$$S \sim \sqrt{N} \propto N_{rel}\,. \tag{3.28}$$

While frozen and chaotic networks have a finite basin entropy, the basin entropy of critical networks increases as the number of relevant nodes [18].

3.4.5
Problems

1. How many cycles does an even (odd) loop of size 6 have?
2. Count the attractors of the network shown in Fig. 3.8 for all four cases where loop 1 and/or loop 2 are even/odd.
3. How does the transient time (i.e. the number of time steps until the network reaches an attractor) increase with N for (a) $K = 1$ networks in the frozen phase, (b) critical $K = 1$ networks, (c) chaotic $K = N$ networks?
4. Consider the subensemble of all critical $K = 1$ networks that have the same wiring, but all possible assignments of "copy" and "invert" functions. Which property determines the probability that a network has a fixed point attractor? If it has such an attractor, how many fixed point attractors does the network have in total? Conclude that there is on an average one fixed point per network in this subensemble.
5. Verify the identity $\bar{k}_l = H_L$ used in calculation (3.25).
6. How does the basin entropy for $K = 1$ networks depend on the parameter δ when δ becomes very small? Find an answer without performing any calculations.

3.5
Critical Networks with *K* = 2

In the previous chapter, we have derived many properties of frozen, critical and chaotic networks by studying ensembles with $K = 1$. Many results are also valid for RBNs with general values of K. In this chapter, we focus on critical $K = 2$ networks. These networks, as well as critical networks with larger values of K, differ in one important respect from critical $K = 1$ networks: they have a *frozen core*, consisting of nodes that are frozen on the same value on all attractors. We have obtained this result already with the annealed approximation: The normalized Hamming distance between two identical networks is close to the critical point given by Eq. (3.12), which means that it is zero exactly at the critical point. For $K = 1$, there exists no chaotic phase and no Equation (3.12), and therefore the observation that all nodes may be nonfrozen in critical $K = 1$ networks is not in contradiction with the annealed approximation.

We will first explain phenomenologically the features of critical $K = 2$ networks, and then we will derive some of these features analytically.

3.5.1 Frozen and Relevant Nodes

The frozen core arises because there are constant functions that fix the values of some nodes, which in turn lead to the fixation of the values of some other nodes, etc. Let us consider Fig. 3.9 as an example. This network has the same number of constant and reversible functions, as is required for critical networks (although this classification only makes sense for large networks, where the thermodynamic limit becomes visible). Node 5 has a constant function and is therefore frozen on the value 0 (indicated by a darker grey shade)

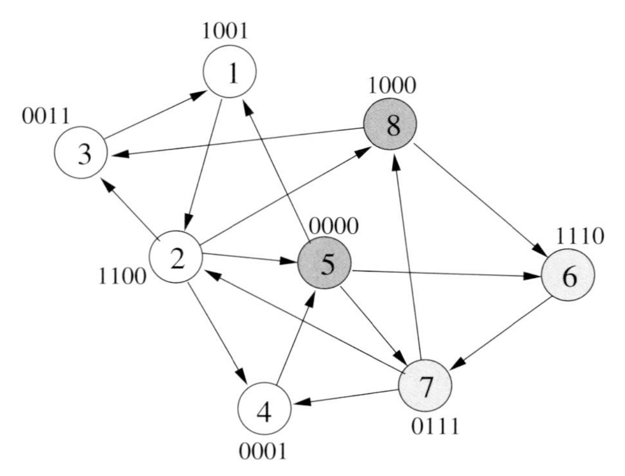

Fig. 3.9 Example of a network with 8 nodes and $K = 2$. The functions are those of Table 3.2 (but written horizontally instead of vertically), with the first input node being the one with the lower number.

after the first time step. Node 6 has a canalyzing function which gives 1 as soon as one of the inputs is 0. Therefore node 6 is frozen in state 1 (indicated by a lighter grey shade) no later than after the second time step. Then node 7 has two frozen inputs and becomes therefore also frozen. Its value is then 1. Node 8 has a canalyzing function which gives 0 as soon as one of the inputs is 1, and will therefore end up in state 0. These four nodes constitute the frozen core of this network. At most after 4 time steps, each of these nodes assumes its stationary value. If we remove the frozen core, we are left with a $K = 1$ network consisting of nodes 1 to 4, with "copy" and "invert" functions between these nodes. For instance, node 4 copies the state of node 2 if node 7 is in state 1. Node 3 copies the state of node 2, node 1 inverts the input it receives from node 3, and node 2 inverts the input it receives from node 1. The nodes 1, 2, 3 form an even loop, and node 4 is slaved to this loop. Nodes 1, 2, 3 are therefore the *relevant nodes* that determine the attractors. We can conclude that this network has 4 attractors: two fixed points and two cycles of length 3.

There is a different mechanism by which a frozen core can arise, which is illustrated by assigning another set of update functions to the same network, as shown in Fig. 3.10. This network contains only canalyzing update functions of the type C_2, and such a network could be classified as critical if it was much larger. We begin again by fixing node 5 at value 1, and we denote this as 5_1. In the next time step, this node may have changed its state, but then node 7 will be in state 1, because it is canalyzed to this value by node 5. By continuing this consideration, we arrive at the following chain of states:

$$5_1 \rightarrow 7_1 \rightarrow 4_1 \rightarrow 5_0 \,.$$

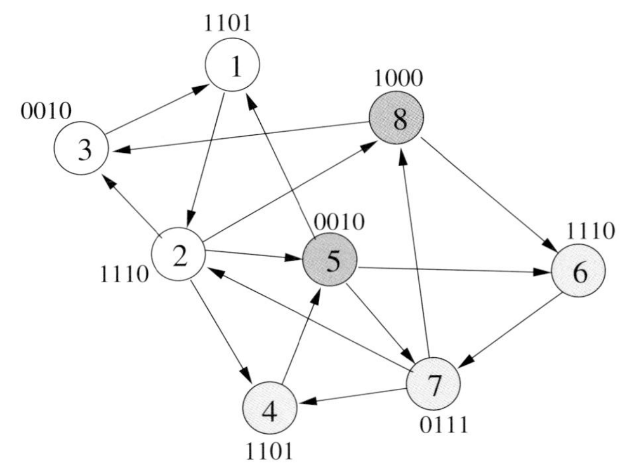

Fig. 3.10 A network with the same topology as the previous network, but with only canalyzing functions.

This means that node 5 must eventually assume the state 0, and we continue from here by following again canalyzing connections:

$$5_0 \to 6_1 \to 7_1 \to (4_1, 8_0) \to (5_0, 6_1) \to (6_1, 7_1) \to (4_1, 7_1, 8_0) \to (4_1, 5_0, 6_1, 8_0)$$
$$\to (5_0, 6_1, 7_1) \to (4_1, 6_1, 7_1, 8_0) \to (4_1, 5_0, 6_1, 7_1, 8_0) \to (4_1, 5_0, 6_1, 7_1, 8_0)$$

From this moment on, nodes 4 to 8 are frozen. Nodes 1, 2, 3 form a relevant loop with the functions invert, invert, copy, just as in the previous example.

In order to better understand how the frozen core arises in this case, consider the loop formed by the nodes 6, 7, 8: This is a *self-freezing loop*. If the nodes 6, 7, 8 are in the states 1, 1, 0, they remain forever in these states, because each node is canalyzed to this value by the input it receives within the loop. This loop has the same effect on the network as have nodes with constant functions. Once this loop is frozen, nodes 4 and 5 become also frozen. One can imagine networks where such a loop never freezes, but this becomes very unlikely for large networks.

The networks shown in the previous two figures were designed to display the desired properties. In general, small networks differ a lot in the number of frozen and nonfrozen nodes, as well as in the size and structure of their relevant component(s) and their attractors. The specific properties particular to the frozen and chaotic phase and to the critical line become clearly visible only for very large networks.

A network of intermediate size is the basis of Fig. 3.11, which shows the nonfrozen part of a critical $K = 2$ network with 1000 nodes. There are 100 nonfrozen nodes in this network, indicating that the majority of nodes are frozen. Among the 100 nonfrozen nodes, only 5 nodes are relevant, and only 6 nodes have two nonfrozen inputs. The relevant nodes are arranged in 2 relevant components. They determine the attractors of the network, while all other nodes sit on outgoing trees and are slaved to the dynamics of the relevant nodes. This figure resembles a lot a $K = 1$ network. The only difference is that there are a few nodes with two inputs.

Analytical calculations, part of which are explained in the next section, give the following general results for critical $K = 2$ networks in the thermodynamic limit $N \to \infty$:

1. The number of nodes that do not belong to the frozen core, is proportional to $N^{2/3}$ for large N.

2. If the proportion of nodes with a constant function is nonzero, the frozen core can be determined by starting from the nodes with constant functions and following the cascade of freezing events.

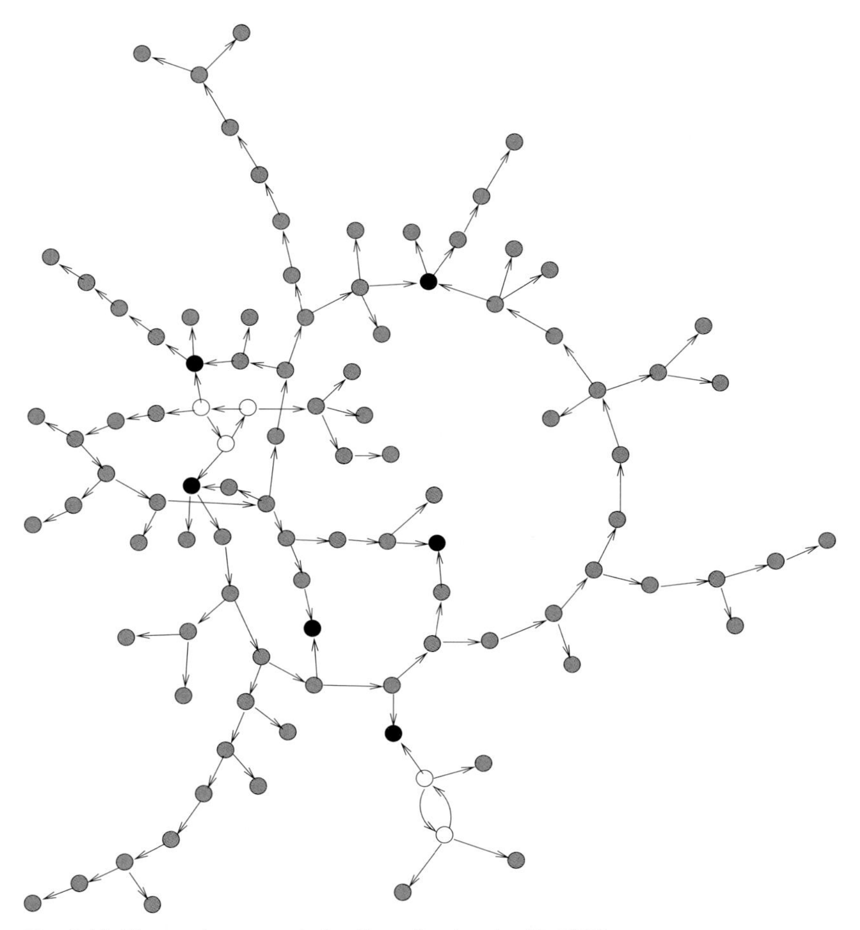

Fig. 3.11 The nonfrozen part of a $K = 2$ network with 1000 nodes. Shown are the 100 nonfrozen nodes. 5 nodes are relevant (white), and 6 nodes (black) have two nonfrozen inputs.

3. If the proportion of nodes with a constant function is zero (which means that the network contains only canalyzing functions), the frozen core can be determined by starting from self-freezing loops.

4. The number of nodes that are nonfrozen and that receive 2 nonfrozen inputs is proportional to $N^{1/3}$.

5. The number of relevant nodes is proportional to $N^{1/3}$. They are connected to relevant components, which consist of loops and possibly additional links within and between the loops.

6. The number of relevant nodes that have two relevant inputs remains finite in the limit $N \to \infty$.

7. The number of relevant components increases as $\log N^{1/3}$.

8. The cutoff of the size of relevant components scales as $N^{1/3}$.

The complete list of these results is given in [19], but part of the results can be found in earlier papers [20–22].

3.5.2
Analytical Calculations

After this qualitative introduction to critical networks, let us derive the main results for the scaling of the number of nonfrozen and relevant nodes with N. Computer simulations of critical networks show the true asymptotic scaling only for very larger networks with more than 100000 nodes. For this reason, the values $2/3$ and $1/3$ for the critical exponents characterizing the number of nonfrozen and relevant nodes has been known only since 2003.

Flyvbjerg [23] was the first one to use a dynamical process that starts from the nodes with constant update functions and determines iteratively the frozen core. Performing a mean-field calculation for this process, he could identify the critical point. We will go now beyond mean-field theory.

We consider the ensemble of all $K = 2$ networks of size N with update rule 2 (weighted functions), where the weights of the $\mathcal{C}_1$, reversible, $\mathcal{C}_2$ and constant functions are α, β, γ and δ. These networks are critical for $\beta = \delta$. We begin by assigning update functions to all nodes and by placing these nodes according to their functions in four containers labelled $\mathcal{F}$, $\mathcal{C}_1$, $\mathcal{C}_2$, and $\mathcal{R}$. These containers then contain N_f, N_{c_1}, N_{c_2}, and N_r nodes. We treat the nodes in container $\mathcal{C}_1$ as nodes with only one input and with the update functions "copy" or "invert". As we determine the frozen core, the contents of the containers will change with time. The "time" we are defining here is not the real time for the dynamics of the system. Instead, it is the time scale for the process that we use to determine the frozen core. One "time step" consists in choosing one node from the container $\mathcal{F}$, in selecting the nodes to which this node is an input, and in determining its effect on these nodes. These nodes change containers accordingly. Then the frozen node need not be considered any more and is removed from the system. The containers now contain together one node less than before. This means that container $\mathcal{F}$ contains only those frozen nodes, the effect of which on the network has not yet been evaluated. The other containers contain those nodes that have not (yet) been identified as frozen. The process ends when container $\mathcal{F}$ is empty (in which case the remaining nodes are the nonfrozen nodes), or when all the other containers are empty (in which case the entire network freezes). The latter case means that the dynamics of the network go to the same fixed point for all initial conditions.

This process is put into the following equations, which describe the changes of the container contents during one "time step".

$$\begin{aligned}
\Delta N_r &= -\frac{2N_r}{N} \\
\Delta N_{c_2} &= -\frac{2N_{c_2}}{N} \\
\Delta N_{c_1} &= \frac{2N_r}{N} + \frac{N_{c_2}}{N} - \frac{N_{c_1}}{N} \\
\Delta N_f &= -1 + \frac{N_{c_2}}{N} + \frac{N_{c_1}}{N} + \xi \\
\Delta N &= -1
\end{aligned} \tag{3.29}$$

The terms in these equations mean the following: Each node in container $\mathcal{R}$ chooses the selected frozen node as an input with probability $2/N$ and becomes then a $\mathcal{C}_1$-node. This explains the first equation and the first term in the third equation. Each node in container $\mathcal{C}_2$ chooses the selected frozen node as an input with probability $2/N$. With probability $1/2$, it then becomes frozen, because the frozen node is with probability $1/2$ in the state that fixes the output of a $\mathcal{C}_2$-node. If the $\mathcal{C}_2$-node does not become frozen, it becomes a $\mathcal{C}_1$-node. This explains the terms proportional to N_{c_2}. Each node in container $\mathcal{C}_1$ chooses the selected frozen node as an input with probability $1/N$. It then becomes a frozen node. Finally, the -1 in the equation for ΔN_f means that the chosen frozen node is removed from the system. In summary, the total number of nodes, N, decreases by one during one time step, since we remove one node from container $\mathcal{F}$. The random variable ξ captures the fluctuations around the mean change ΔN_f. It has zero mean and variance $(N_{c_1} + N_{c_2})/N$. The first three equations should contain similar noise terms, but since the final number of nodes of each class is large for large N, the noise can be neglected in these equations. We shall see below that at the end of the process most of the remaining nodes are in container $\mathcal{C}_1$, with the proportion of nodes left in containers $\mathcal{C}_2$ and $\mathcal{R}$ vanishing in the thermodynamic limit. Figure 3.12 illustrates the process of determining the frozen core.

The number of nodes in the containers, N, can be used instead of the time variable, since it decreases by one during each step. The equations for N_r and N_{c_2} can then be solved by going from a difference equation to a differential equation,

$$\frac{\Delta N_r}{\Delta N} \simeq \frac{dN_r}{dN} = -\frac{2N_r}{N},$$

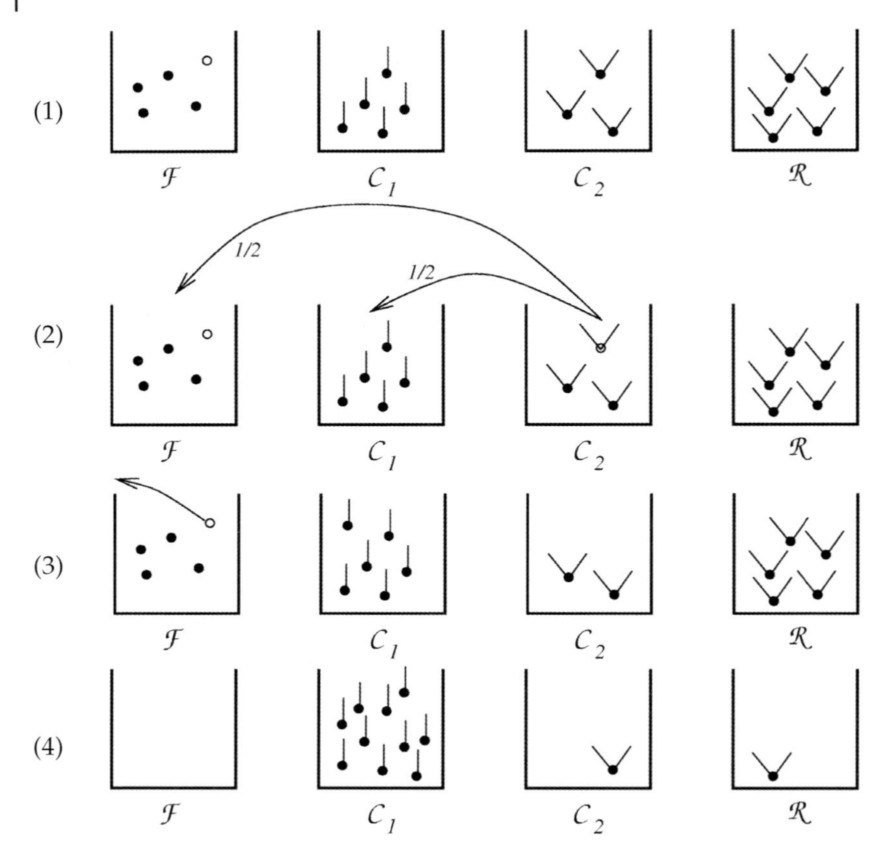

Fig. 3.12 Illustration of the freezing process. (1) Initially, a frozen node is chosen (marked in white), (2) then it is determined to which node(s) this is an input and the effect on those nodes is determined. (3) Then, the selected frozen node is removed. (4) The last picture sketches the final state, where all frozen nodes have been removed and most remaining nodes have 1 nonfrozen input.

which has the solution

$$N_r = \frac{\beta N^2}{N^{ini}}, \qquad N_{c_2} = \frac{\gamma N^2}{N^{ini}}, \tag{3.30}$$

where we have now denoted the total number of nodes with N^{ini}, since the value of N changes during the process. Similarly, we find if we neglect the noise term for a moment

$$N_f = N(\delta - \beta) + \frac{\beta N^2}{N^{ini}}, \qquad N_{c_1} = N(\alpha + \gamma + 2\beta) - 2\frac{N^2(\beta + \gamma)}{N^{ini}}. \tag{3.31}$$

From this result, one can derive again the phase diagram, as we did by using the annealed approximation: For $\delta < \beta$, i.e. if there are more frozen than reversible update functions in the network, we obtain $N_f = 0$ at a nonzero value of N, and the number of nonfrozen nodes is proportional to N^{ini}. We are in the chaotic phase. For $\delta > \beta$, there exists no solution with $N_f = 0$ and $N > 0$. The network is in the frozen phase. For the critical networks that we want to focus on, we have $\delta = \beta$, and the process stops at $N_f = 1 = \beta N^2/N^{ini}$ if we neglect noise. This means that $N = \sqrt{N^{ini}/\beta}$ at the end of the process. The number of nonfrozen nodes would scale with the square root of the network size. This is not what is found in numerical studies of sufficiently large networks. We therefore must include the noise term. Noise becomes important only after N_f has become small, when most nodes are found in container C_1, and when the variance of the noise has become unity, $\langle \xi^2 \rangle = 1$. Inserting the solution for N_r into the equation for N_f, we obtain then

$$\frac{dN_f}{dN} = \frac{N_f}{N} + \frac{\beta N}{N^{ini}} + \xi \tag{3.32}$$

with the step size $dN = 1$. We want to transform this into a Fokker–Planck-equation. Let $P(N_f, N)$ be the probability that there are N_f nodes in container $\mathcal{F}$ at the moment where there are N nodes in total in the containers. This probability depends on the initial node number N_{ini}, and on the parameter β. The sum

$$\sum_{N_f=1}^{\infty} P(N_f, N) \simeq \int_0^{\infty} P(N_f, N) dN_f$$

is the probability that the stochastic process is not yet finished, i.e. the probability that N_f has not yet reached the value 0 at the moment where the total number of nodes in the containers has decreased to the value N. This means that systems that have reached $N_f = 0$ must be removed from the ensemble, and we therefore have to impose the absorbing boundary condition $P(0, N) = 0$. Exactly in the same way as with calculation (3.14), we obtain then

$$-\frac{\partial P}{\partial N} = \frac{\partial}{\partial N_f}\left(\frac{N_f}{N} + \frac{\beta N}{N^{ini}}\right) P + \frac{1}{2}\frac{\partial^2 P}{\partial N_f^2}. \tag{3.33}$$

We introduce the variables

$$x = \frac{N_f}{\sqrt{N}} \text{ and } y = \frac{N}{(N^{ini}/\beta)^{2/3}} \tag{3.34}$$

and the function $f(x, y) = (N^{ini}/\beta)^{1/3} P(N_f, N)$. We will see in a moment that $f(x, y)$ does not depend explicitely on the parameters N^{ini} and β with this definition. The Fokker–Planck equation then becomes

$$y\frac{\partial f}{\partial y} + f + \left(\frac{x}{2} + y^{3/2}\right)\frac{\partial f}{\partial x} + \frac{1}{2}\frac{\partial^2 f}{\partial x^2} = 0\,. \tag{3.35}$$

Let $W(N)$ denote the probability that N nodes are left at the moment where N_f reaches the value zero. It is

$$W(N) = \int_0^\infty P(N_f, N)\, dN_f - \int_0^\infty P(N_f, N-1)\, dN_f\,.$$

Consequently,

$$\begin{aligned} W(N) &= \frac{\partial}{\partial N}\int_0^\infty P(N_f, N) dN_f \\ &= (N^{ini}/\beta)^{-1/3}\frac{\partial}{\partial N}\sqrt{N}\int_0^\infty f(x,y)\, dx \\ &= (N^{ini}/\beta)^{-2/3}\frac{\partial}{\partial y}\sqrt{y}\int_0^\infty f(x,y)\, dx \\ &\equiv (N^{ini}/\beta)^{-2/3} G(y) \end{aligned} \tag{3.36}$$

with a scaling function $G(y)$. $W(N)$ must be a normalized function,

$$\int_0^\infty W(N)\, dN = \int_0^\infty G(y)\, dy = 1\,.$$

This condition is independent of the parameters of the model, and therefore $G(y)$ and $f(x,y)$ are independent of them, too, which justifies our choice of the prefactor in the definition of $f(x,y)$. The mean number of nonfrozen nodes is therefore

$$\bar{N} = \int_0^\infty N W(N)\, dN = (N^{ini}/\beta)^{2/3}\int_0^\infty G(y) y\, dy\,, \tag{3.37}$$

which is proportional to $(N^{ini}/\beta)^{2/3}$. From Eqs. (3.30) and the corresponding equation for the C_2-nodes we find then that the number of nonfrozen nodes with two nonfrozen inputs is proportional to $N^{1/3}$. This is a vanishing proportion of all nonfrozen nodes.

The nonfrozen nodes receive their (remaining) input from each other, and we obtain the nonfrozen part of the network by randomly making the remaining connections. If we neglect for a moment the second input of those nonfrozen nodes that have two nonfrozen inputs, we obtain a $K = 1$ network. The number of relevant nodes must therefore be proportional to the square root of number of nonfrozen nodes, i.e. it is $N_{rel} \sim N^{1/3}$, and the number of relevant components is of the order $\ln N^{1/3}$, with the largest component of the order of $N^{2/3}$ nodes (including the trees). Adding the second input to the nonfrozen nodes with two nonfrozen inputs does not change much: The total number of relevant nodes that receive a second input is a constant (since each of $\sim N^{1/3}$ relevant nodes receives a second input with a probability proportional to $N^{-1/3}$). Only the largest loops are likely to be affected, and therefore only the large relevant components may have a structure that is more complex than a simple loop. Most nonfrozen nodes with two nonfrozen inputs sit in the trees, as we have seen in Fig. 3.11. The mean number and length of attractors can now be estimated in the following way: The attractor number must be at least as large as the number of cycles on the largest relevant loop, and therefore it increases exponentially with the number of relevant nodes. The mean attractor length becomes larger as for $K = 1$ networks, since complex relevant components can have attractors that comprise a large part of their state space, as was shown in [24]. Such components arise with a nonvanishing probability, and they dominate therefore the mean attractor length, which therefore increases now exponentially with the number of relevant nodes.

The conclusions derived in the last paragraph can be made more precise. Interested readers are referred to [19].

All these results are also valid for $K = 2$ networks with only canalyzing functions. As mentioned before, the frozen core of canalyzing networks arises through self-freezing loops. The resulting power laws are the same as for networks with constant functions, as was shown in [25].

3.5.3
Problems

1. What is the number of attractors of the network shown in Fig. 3.11 for all four cases where loop 1 and/or loop 2 are even/odd?
2. Assume there are 4 relevant nodes, one of them with two relevant inputs. List all topologically different possibilities for the relevant components.
3. Using Eq. (3.36), figure out how the probability that the entire network freezes depends on N.

3.6
Networks with Larger *K*

Just as we did for $K = 2$, we consider larger values of K only for those update rules that lead to fixed points of b_t (i.e. of the proportion of 1s), and therefore to a critical line separating a frozen and a chaotic phase.

Let us first consider the frozen phase, where the sensitivity λ is smaller than 1. The probability that a certain node is part of a relevant loop of size

l is for large N obtained by the following calculation: the node has K inputs, which have again each K inputs, etc., so that there are K^{l-1} nodes that might choose the first node as one of its K inputs, leading to a *connection loop*. The chosen node is therefore part of K^l/N connection loops of length l on an average. The probability that a given connection loop has no frozen connection is $(\lambda/K)^l$, and therefore the mean number of relevant loops of size l is λ^l/l. The mean number of relevant nodes is then

$$\langle N_{rel} \rangle = \sum_l \lambda^l = \frac{\lambda}{1-\lambda} . \tag{3.38}$$

This is the same result as Eq. (3.22), which we derived for $K = 1$. The mean number of nonrelevant nodes to which a change of the state of a relevant node propagates is given by the same sum, since in each step the change propagates on an average to λ nodes. By adding the numbers of relevant and nonrelevant nonfrozen nodes, we therefore obtain again a mean number of $\lambda/(1-\lambda)^2$ nonfrozen nodes, just as in the case $K = 1$. We conclude that the frozen phases of all RBNs are very similar.

Now we consider critical networks with $K > 2$. The number of nonfrozen nodes scales again as $N^{2/3}$ and the number of relevant nodes as $N^{1/3}$. The number of nonfrozen nodes with k nonfrozen inputs scales with N as $N^{(3-k)/3}$. These results are obtained by generalizing the procedure used in the previous chapter for determining the frozen core [26]. By repeating the considerations of the previous paragraph with the value $\lambda = 1$, we find that in all critical networks the mean number of relevant loops of size l is $1/l$ – as long as l is smaller than a cutoff, the value of which depends on N. For $K = 1$ the cutoff is at $\sqrt{N}$, for $K = 2$, it is at $N^{1/3}$, and this value does not change for larger K. There exists a nice phenomenological argument to derive the scaling $\sim N^{2/3}$ of the number of nonfrozen nodes [27]: The number of nonfrozen nodes should scale in the same way as the size of the largest perturbation, since the largest perturbation affects all nodes on the largest nonfrozen component. The cutoff s_{max} in the size of perturbations (see Eq. (3.13)) is given by the condition that $n(s_{max}) \sim 1/N$. Perturbations larger than this size occur only rarely in networks of size N, since $n(s)$ is the probability that a perturbation of one specific node (out of N the nodes) affects s nodes in total. Using Eq. (3.13), we therefore obtain

$$s_{max} \sim N^{2/3} . \tag{3.39}$$

This argument does not work for $K = 1$, where we have obtained $s_{max} \sim N$ in Section 3.4. The reason is that critical networks with $K = 1$ have no frozen core, but every node that receives its input from a perturbed node will also be perturbed.

As far as the chaotic phase is concerned, there are good reasons to assume that it displays similar features for all K. We have explicitely considered the case $K = N$. Numerical studies show that the basin entropy approaches a constant with increasing K also when the value of K is fixed [18]. When λ is close to 1, there is a frozen core that comprises a considerable part of the network. We can expect that the nonfrozen part has a state space structure similar to that of the $K = N$ networks.

3.7 Outlook

There are many possibilities of how to go beyond RBNs with synchronous update. In this last chapter, we will briefly discuss some of these directions.

3.7.1 Noise

Synchronous update is unrealistic since networks do not usually have a central pacemaker that tells all nodes when to perform the next update. Asynchronous update can be done either deterministically by assigning to each node an update time interval and an initial phase (i.e. the time until the first update), or stochastically by assigning to each node a time-dependent probability for being updated. We focus here on stochastic update, since all physical systems contain some degree of noise. In particular, noise is ubiquitous in gene regulatory networks [28]. Boolean networks with stochastic update are for instance investigated in [29, 30]. The frozen core obviously remains a frozen core under stochastic update, and the relevant nodes remain relevant nodes. The most fundamental change that occurs when one switches from deterministic to stochastic update is that there is now in general more than one successor to a state. The set of recurrent states comprises those states that can reoccur infinitely often after they have occurred for the first time. However, if there is a path in state space from each state to a fixed point or to a cycle that has only one successor for each state, the network behaves deterministically for large times, in spite of the stochastic update. This occurs in networks where all relevant nodes sit on loops: an even loop has two fixed points, and an odd loop has an attractor cycle of length $2l$, where each state has only one successor in state space (apart from itself). If the number of relevant loops increases logarithmically with system size N, the number of attractors then increases as a power law of N. This means that critical $K = 1$ networks with asynchronous update have attractor numbers that increases as a power law with system size. In [30] it is argued that in critical $K = 2$ networks, where not all relevant components are simple loops, the attractor number is still a power law in N.

The situation becomes different when the noise does not only affect the update time but also the update function. Then the output of a node can deviate from the value prescribed by the update function with a probability that depends on the strength of the noise. The interesting question to address in this context is whether the networks remain in the neighborhood of one attractor (which can be tested by evaluating the return probability after switching off the noise), or whether they move through large regions of state space. Investigations of networks with such a type of noise can be found in [31,32].

3.7.2
Scale-free Networks and Other Realistic Network Structures

Real networks do not have a fixed number of inputs per node, but do often have a power-law distribution in the number of inputs or the number of outputs [33]. Boolean dynamics on such networks has been studied [34], however, how this affects the power laws in critical networks, is only partially known [27].

There are many more characteristics of real networks that are not found in random network topologies, such as clustering, modularity, or scale invariance. The effect of all these features on the network dynamics is not yet sufficiently explored.

3.7.3
External Inputs

Real networks usually have some nodes that respond to external inputs. Such an external input to a node can be modelled by switching the constant function from 1 to 0 or vice versa. The set of nodes that cannot be controlled in this way is called the computational core. Networks with a higher proportion of C_2 functions tend to have a larger computational core, since the C_2 functions can mutually fix or control each other. Investigations of this type can be found in [35].

3.7.4
Evolution of Boolean Networks

Ensembles of networks that are completely different from the random ensembles studied in this review can be generated by evolving networks using some rule for mutations and for the network "fitness". For instance, by selecting for robustness of the attractors under noise, one obtains networks with short attractors that have large basins of attraction, but that do not necessarily have a large frozen core [36–38].

In another class of evolutionary models, fitness is not assigned to the entire network, but links or functions are changed if they are associated with nodes that do not show the "desired" behavior, for instance if they are mostly frozen (or active), or if they behave most of the time like the majority of other nodes [39–41].

3.7.5
Beyond the Boolean Approximation

There exist several examples of real networks, where the essential dynamical steps can be recovered when using simple Boolean dynamics. If a sequence of states shall be repeatable and stable, and if each state is well enough approximated by an "on"-"off" description for each node, Boolean dynamics should be a good approximation. However, wherever the degree of activity of the nodes is important, the Boolean approximation is not sufficent. This is the case for functions such as continuous regulation or stochastic switching or signal amplification. Clearly, in those cases a modelling is needed that works with continuous update functions or rate equations based on concentrations of molecules. The different types of network modelling are reviewed for instance in [42].

Acknowledgement

Many results reported in this article were obtained in collaboration with former and present members of my group, Viktor Kaufman, Tamara Mihalev, Florian Greil, Agnes Szejka. I want to thank all those who read earlier versions of this article and made suggestions for improvements: Tamara Mihalev, Florian Greil, Agnes Szejka, Christoph Hamer, Carsten Marr, Christoph Fretter.

References

1 S. A. Kauffman, *J. Theor. Biol.* 22 **(1969)**, p. 437.

2 S. A. Kauffman, *Nature* 224 **(1969)**, p. 177.

3 M. Aldana-Gonzalez, S. Coppersmith, and L. P. Kadanoff, in *Perspectives and Problems in Nonlinear Science, A celebratory volume in honor of Lawrence Sirovich*, eds. E. Kaplan, J. E. Mardsen and K. R. Sreenivasan (**2003**), Springer, New York.

4 S. E. Harris et al., *Complexity* 7 (**2002**), p. 23.

5 S. A. Kauffman, C. Peterson, B. Samuelsson, and C. Troein, *Proc. Nat. Acad. Sci. USA* 101 (**2004**), p. 17102.

6 A. A. Moreira and L. A. N. Amaral, *Phys. Rev. Lett.* 94 (**2005**), 218702.

7 F. Li, T. Long, Y. Lu, Q. Quyang, and C. Tang, *Proc. Nat. Acad. Sci. USA* 101 (**2004**), p. 4781.

8 T. Rohlf and S. Bornholdt, *Physica A* 310 (**2002**), p. 245.

9 S. Bornholdt and K. Sneppen, *Proc. Roy. Soc. Lond. B* 267 (**2000**), p. 2281.

10 B. Samuelsson and C. Troein, *Phys. Rev. Lett.* 90 (**2003**), 098701.

11 B. Derrida and Y. Pomeau, *Europhys. Lett.* 1 (**1986**), p. 45.

12 M. Andrecut and M. K. Ali, *Int. J. Mod. Phys. B* 15 (**2001**), p. 17.

13 I. Shmulevich and S. A. Kauffman, *Phys. Rev. Lett.* 93 (**2004**), 048701.

14 H. Flyvbjerg and N. J. Kjaer, *J. Phys. A* 21 (**1988**), p. 1695.

15 B. Samuelsson and C. Troein, *Phys. Rev. E* 72 (**2005**), 046112.

16 B. Drossel, *Phys. Rev. E* 72 (**2005**), 016110.

17 B. Drossel, T. Mihaljev and F. Greil, *Phys. Rev. Lett.* 94 (**2005**), 088701.

18 P. Krawitz and I. Shmulevich, *Phys. Rev. Lett.* 98 (**2007**), 158701.

19 V. Kaufman, T. Mihaljev and B. Drossel, *Phys. Rev. E* 72 (**2005**), 046124.

20 U. Bastolla and G. Parisi, *Physica D* 115 (**1998**), p. 203.

21 U. Bastolla and G. Parisi, *Physica D* 115 (**1998**), p. 219.

22 J. E. S. Socolar and S. A. Kauffman, *Phys. Rev. Lett.* 90 (**2003**), 068702.

23 H. Flyvbjerg, *J. Phys. A* 21 (**1988**), p. L955.

24 V. Kaufman and B. Drossel, *Eur. Phys. J. B* 43 (**2005**), p. 115.

25 U. Paul, V. Kaufman and B. Drossel,*Phys. Rev. E* 73 (**2006**), 028118.

26 T. Mihaljev and B. Drossel, *Phys. Rev. E* 74 (**2006**), 046101.

27 D.-S Lee and H. Rieger, cond-mat/0605725 (**2006**)

28 H.H. McAdams and A. Arkin, *Proc. Nat. Acad. Sci. USA* 94 (**1997**), p. 814.

29 K. Klemm and S. Bornholdt, *Phys. Rev. E* 72 (**2005**), 055101(R).

30 F. Greil and B. Drossel, *Phys. Rev. Lett.* 95 (**2005**), 048701.

31 I. Shmulevich, E.R. Dougherty, and W. Zhang, *Proceedings of the IEE* 90 ((**2002**), p. 1778.

32 X. Qu, M. Aldana and L. P. Kadanoff, nlin.AO/0207016 (**2006**).

33 R. Albert and A. L. Barabasi, *Rev. Mod. Phys.* 74 (**2002**), p. 47.

34 M. Aldana, *Physica D* 185 (**2003**), p. 45.

35 L. Correale, M. Leone, A. Pagnani, M. Weigt, R. Zecchina, *J. Stat. Mech.* (**2006**), P03002.

36 S. Bornholdt and K. Sneppen, *Phys. Rev. Lett.* 81 (**1998**), p. 236.

37 S. Bornholdt and K. Sneppen, *Proc. Roy. Soc. Lond. B* 267 (**2000**), p. 2281.

38 A. Szejka and B. Drossel, *Eur. Phys. J. B* 56 (**2007**), p. 373.

39 M. Paczuski, K. E. Bassler, and Á. Corral, *Phys. Rev. Lett* 84, (**2000**), p. 3185.

40 K. E. Bassler, C. Lee, and Y. Lee, *Phys. Rev. Lett.* 93 (**2004**), 038101.

41 M. Liu and K. E. Bassler, *Phys. Rev. E* 74 (**2006**), 041910).

42 H. de Jong, *Journal of Computational Biology* 9 (**2002**), p. 67.

4
Return Intervals and Extreme Events in Persistent Time Series with Applications to Climate and Seismic Records

Armin Bunde, Jan F. Eichner, Shlomo Havlin, Jan W. Kantelhardt, and Sabine Lennartz

Abstract

Many natural records exhibit long-term persistence characterized by correlation functions $C(s) \sim s^{-\gamma}$ with exponents γ between 0 and 1. We review our studies of the statistics of return intervals and extreme events (maxima) in long correlated data sets with different (Gaussian, exponential, power-law, and log-normal) distributions and discuss applications to geophysical records. We found (i) a stretched exponential distribution of the return intervals (Weibull distribution with an exponent equal to γ), (ii) clustering of both small and large return intervals, and (iii) a counter-intuitive behavior of the mean residual time to the next extreme event. For maxima within time segments of fixed duration R we found that (i) the integrated distribution function converges to a Gumbel distribution for large R, (ii) the speed of the convergence depends on both, the long-term correlations and the initial distribution of the values, (iii) the maxima series exhibit long-term correlations similar to those of the original data, and most notably (iv) the maxima distribution as well as the mean maxima significantly depend on the history, in particular on the previous maximum. Most of the effects revealed in artificial data are found in hydrological and climatological data series. Finally, we present indications that the magnitudes of earthquakes, in regimes of stationary seismicity, are also long-term correlated, and that the anomalous statistics of their return intervals can be explained by these correlations.

4.1
Introduction

In recent years there is growing evidence that many natural records exhibit long-term correlations [1, 2]. Prominent examples include hydrolog-

Review of Nonlinear Dynamics and Complexity. Edited by Heinz Georg Schuster

ISBN: 978-3-527-40729-3

ical data [3–7], meteorological and climatological records [8–12], turbulence data [13, 14], as well as DNA sequences [15–17], and physiological records [18,19]. Long-term persistence has also been found in the volatility of economic records [20,21] and the internet-traffic [22,23].

In long-term persistent records (x_i), $i = 1, \ldots, N$ with mean $\bar{x}$ and standard deviation σ_x the autocorrelation function $C_x(s)$ decays by a power law,

$$\begin{aligned} C_x(s) &= \frac{1}{\sigma_x^2} \langle (x_i - \bar{x})(x_{i+s} - \bar{x}) \rangle \\ &\equiv \frac{1}{\sigma_x^2 (N-s)} \sum_{i=1}^{N-s} (x_i - \bar{x})(x_{i+s} - \bar{x}) \sim s^{-\gamma} \end{aligned} \tag{4.1}$$

where γ denotes the correlation exponent, $0 < \gamma < 1$. Such correlations are named 'long-term' since the mean correlation time $T = \int_0^\infty C_x(s)\, ds$ diverges for infinitely long series (in the limit $N \to \infty$). Power-law long-term correlations according to Eq. (4.1) correspond to a power spectrum $P(f) \sim f^{-\beta}$ with $\beta = 1 - \gamma$ (Wiener–Khintchin theorem).

Figure 4.1 shows three examples with parts of annual data series exhibiting different degrees of long-term correlations. These records are (a) the reconstructed river run-off of the Sacramento River (USA) [24], (c) the historical water level minima of the Nile River (EGY) [25], and (e) the reconstructed northern hemisphere temperature [26]. The figure also shows the histograms $H(x)$ of the values x_i of the data; we denote the corresponding distribution density by $P(x)$. To test the records for long-term correlations, we have employed the 2nd order detrended fluctuation analysis (DFA2) [15,18,30]. In general, in DFAn, one considers the cumulated sum of the data $Y_j = \sum_{i=1}^{j} x_i$ and studies, in time windows of length s, the mean fluctuation $F(s)$ of Y_j around the best polynomial fit of order n, i.e. in DFA2 around the best quadratic fit. For long-term correlated data, $F(s)$ scales as $F(s) \sim s^{\alpha}$, with $\alpha = 1 - \gamma/2$. To standardize the records, we subtracted each record by its mean and divided by its variance. The DFA2 fluctuation functions for the three records from Fig. 4.1(a,c,e) are shown in Fig. 4.1(g) together with those obtained from the reconstructed temperature record of Baffin Island [27] and the reconstructed precipitation in New Mexico (USA) [28]. The measured γ values are 0.8 (Sacramento), 0.55 (Baffin), 0.4 (New Mexico), 0.3 (Nile), and 0.1 (Northern hemisphere). The observed phenomena should also occur in heartbeat records, internet traffic and stock market volatility and have to be taken into account for an efficient risk evaluation.

In the following we describe how the presence of long-term correlations in general affects the statistics of the return intervals r between events above a certain threshold value q [29, 31–33]. We show that the long-term memory in the data leads to a stretched exponential distribution of long return inter-

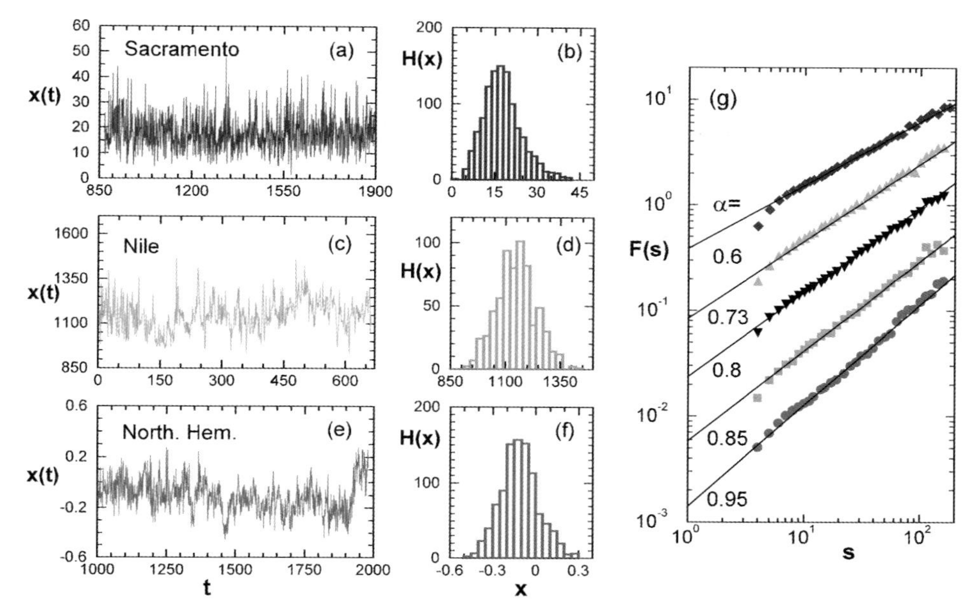

Fig. 4.1 Three annual data series $x(t)$ and their distribution histograms $H(x)$; (a, b) reconstructed river run-off of the Sacramento River (USA) [24], (c, d) historical water level minima of the Nile River (EGY) [25], and (e, f) reconstructed northern hemisphere temperature [26]. The data and more information about it can be achieved from the NOAA Paleoclimatology Program on the internet: http://www.ngdc.noaa.gov/paleo/recons.html. In (g) the corresponding correlation exponents have been determined. The slopes α of the DFA2 fluctuation function are related to the γ values via $\alpha = 1 - \gamma/2$; the results for the reconstructed temperature record of Baffin Island [27] and the reconstructed precipitation in New Mexico (USA) [28] are also included. The measured γ values are 0.8 (Sacramento), 0.55 (Baffin), 0.4 (New Mexico), 0.3 (Nile), and 0.1 (Northern hemisphere). (Adapted from [29]).

vals, $\ln P_q(r) \sim -r^\gamma$, complemented by a power-law behavior for short return intervals. In short long-term correlated records, however, the stretched exponential can become hard to be observed and to be differentiated from a simple exponential decay. In addition, we have found long-term correlations among the return intervals themselves, yielding pronounced clustering of both small and large return intervals. Classical autocorrelation analysis is more suitable for the study of these correlation properties of return interval series than sophisticated DFA approaches. Finally, we observed an anomalous behavior of the mean residual time to the next event that depends on the history and increases with the elapsed time in a counterintuitive way. We present an analytical scaling approach and demonstrate that all these features can be seen in long climate records.

The presence of long-term correlations also affects the statistics of the extreme events, i.e., the maxima values of the signal within time segments of fixed duration R [34]. We have found numerically that (1) the integrated distribution function of the maxima converges to a Gumbel distribution for large R similar to uncorrelated signals, (2) the deviations for finite R depend on the initial distribution of the records and on their correlation properties, (3) the maxima series exhibit long-term correlations similar to those of the original data, and most notably (4) the maxima distribution as well as the mean maxima significantly depend on the history, in particular on the previous maximum. The last item implies that conditional mean maxima and conditional maxima distributions (with the value of the previous maximum as condition) should be considered for improved extreme event prediction. We provide indications that this dependence of the mean maxima on the previous maximum occurs also in observational long-term correlated records.

Finally, we study as a particular example, seismic records in regimes of stationary seismic activity in Northern and Southern California [35]. We have found that the magnitudes of the earthquakes are long-term correlated with a correlation exponent around $\gamma = 0.4$. We show explicitly that these long-term correlations can explain the correlations among the return intervals between events above a certain magnitude M and, without any fit parameter, the scaling form of the distribution function of the return intervals in the seismic records, recently obtained by Corral [36].

4.2 Statistics of Return Intervals

The statistics of return intervals between extreme events is a powerful tool to characterize the temporal scaling properties of experimental time series and to derive quantities for the estimation of the risk for hazardous events like floods, very high temperatures, or earthquakes. We focus on long-term correlated signals with different distributions of the values (Gaussian, exponential, power-law, and log-normal distributions) [31].

4.2.1 Data Generation and Mean Return Interval

In our numerical procedure, we generated sequences of random numbers (x_i) of length $N = 2^{21}$ with either Gaussian, exponential, power-law, or log-normal distribution with unit variance. The corresponding distribution densities $P(x)$ are given by:

$$P_{\text{Gauss}}(x) = \frac{1}{\sqrt{2\pi\sigma}} \exp\left(-x^2 \Big/ 2\sigma\right) \tag{4.2a}$$

$$P_{\exp}(x) = \frac{1}{x_0} \exp\left(-x/x_0\right) \tag{4.2b}$$

$$P_{\text{power}}(x) = (\delta - 1)\, x^{-\delta} \tag{4.2c}$$

$$P_{\text{log-norm}}(x) = \frac{1}{\sqrt{2\pi}x} \exp\left[-(\ln x + \mu)^2/2\right] \tag{4.2d}$$

We chose, without loss of generality, $\sigma = 1$, $x_0 = 1$, and $\mu = 0.763$ in Eqs. (4.2a), (4.2b), (4.2d) and select $\delta = 5.5$ in Eq. (4.2c). The long-term correlations were introduced by the Fourier-filtering technique (see, e.g. [37,38]). To preserve the shape of the distribution after retransforming the data from Fourier space we applied an iterative method [39]. For each distribution density $P(x)$ we generated 150 data sets using 1000 iterations, restoring the desired power spectrum by Fourier-filtering and restoring the desired distribution by rank-ordered replacement of the values in each iteration until convergence is achieved. A full explanation of this iterative procedure is given in the appendix of [31]. For the Gaussian data, one iteration is sufficient since Fourier-filtering preserves the Gaussian distribution. We tested the quality of the long-term correlations of the data with detrended fluctuation analysis (DFA) [15,18,30] and autocorrelation function analysis.

For describing the recurrence of rare events exceeding a certain threshold q, defined in units of the standard deviations of the original distribution $P(x)$, we investigated the statistics of the return intervals r between these events as illustrated in Fig. 4.2(a); see also [40]. Figure 4.2(b) shows a section of the sequence of the return intervals for Gaussian long-term correlated data for a threshold q (quantile) chosen such that the mean return interval R_q ('return period') is approximately 100. Figure 4.2(c) shows the same section, but the original data was shuffled before, destroying the correlations. One can see that there are more large r-values and many more short r-values in Fig. 4.2(b) compared to the uncorrelated case in Fig. 4.2(c), although the mean return interval R_q is the same. The long and short return intervals in Fig. 4.2(b) appear in clusters [29], creating epochs of cumulated extreme events caused by the short r-values, and also long epochs of few extreme events caused by the long r-values. In the following we show, how R_q and the distribution density $P_q(r)$ of the return intervals are affected by the presence of long-term correlations as well as by different distribution densities $P(x)$ of the data.

Let us consider the mean return interval R_q. For a given threshold q, there exist N_q return intervals r_j, $j = 1, 2, \ldots, N_q$, which satisfy the sum rule $\sum_{j=1}^{N_q} r_j = N$. When the data are shuffled, the long-term correlations are destroyed, but the sum rule still applies with the same value of N_q. Accordingly, for both long-term correlated and uncorrelated records, the mean return interval $R_q = N/N_q$ is not affected by the long-term correlations. This statement can also be considered as the time series analogous of Kac's Lemma [41].

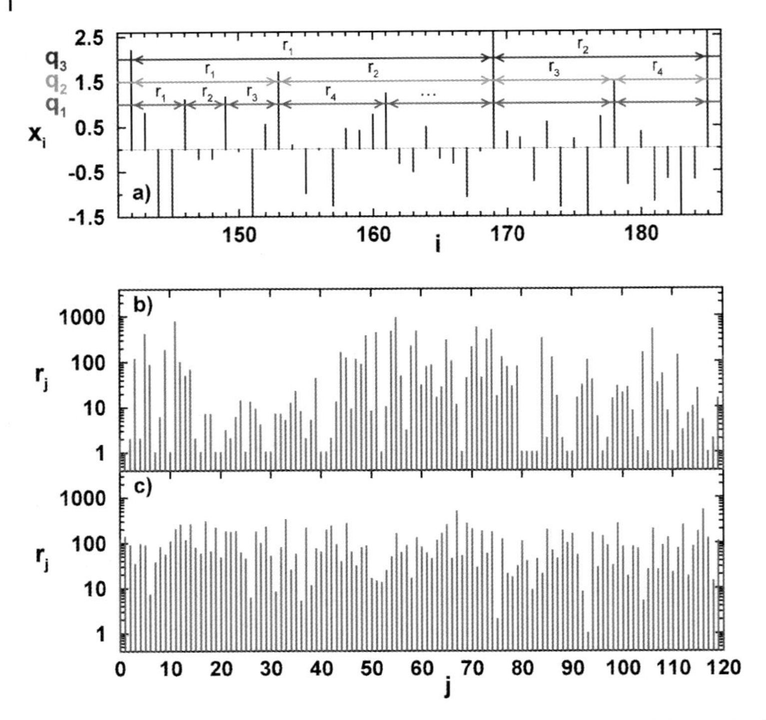

Fig. 4.2 Return intervals: (a) Return intervals r between events x_i above given thresholds q. (b) Sequence of return intervals r of long-term correlated data x_i with $\gamma = 0.4$ for a fixed threshold $q = 2.327$ so that the average r-value is $R_q = 100$. Note the logarithmic scale of the r axis. (c) Sequence of return intervals for uncorrelated data (shuffled x_i of (b)). In (b) more epochs with r-values of small and large size appear compared to (c). (Adapted from [31]).

Hence, R_q can be obtained directly from the tail of the (normalized) distribution density $P(x)$. Accordingly, there is a one-by-one correspondence between q and R_q, which depends only on the distribution density $P(x)$ but not on the correlations.

4.2.2
Stretched Exponential Behavior and Finite-Size Effects for Large Return Intervals

It is known that for uncorrelated records ('white noise'), the return intervals are also uncorrelated and (according to the Poisson statistics) exponentially distributed with $P_q(r) = \frac{1}{R_q} \exp(-r/R_q)$, where R_q is the mean return interval for the given threshold q (see, e.g., [42]). When introducing long-term correlations in Gaussian data the shape of $P_q(r)$ for large values of r ($r > R_q$)

is changed to a stretched exponential [29,31–33,40]

$$P_q(r) \cong \frac{a_\gamma}{R_q} \exp\left[-b_\gamma \left(r/R_q\right)^\gamma\right], \tag{4.3}$$

where the exponent γ is the correlation exponent, and the parameters a_γ and b_γ are independent of q. Their dependence upon γ can be determined from two normalization conditions. Hence, if γ is determined independently by correlation analysis, we can obtain a data collapse of all curves for different values of q by plotting $R_q P_q(r)$ versus r/R_q (according to Eq. (4.2)) [29,32,33]. However, we have to note that Eq. (4.2) does not hold for small values of r, causing some deviations in the values of parameters a_γ and b_γ and the data collapse.

Figure 4.3(a) displays the distribution densities $P_q(r)$ of the return intervals for long-term correlated ($\gamma = 0.4$) and uncorrelated Gaussian data, for 3 values of R_q, $R_q = 15, 44$, and 161. Figure 4.3(b) shows that there exists a data collapse for these 3 curves, when the axis are properly scaled with the return period R_q. This scaling feature is very important since it allows making predictions also for large thresholds q where the statistics is poor. The continuous line in the figure is stretched exponential with $\gamma = 0.4$, as suggested by Eq. (4.2). Figure 4.3(c) further emphasized the stretched exponential behavior. In the double-logarithmic plot of $-\ln[P_q(r)/P_q(1)]$ versus r/R_q, the slopes of the lines are identical to the correlation exponents γ.

Figure 4.4(a–d) shows the rescaled distribution density function of the return intervals for one value of $R_q, R_q = 100$, for the four different types of distributions of the original data (Gaussian, exponential, power-law, and log-

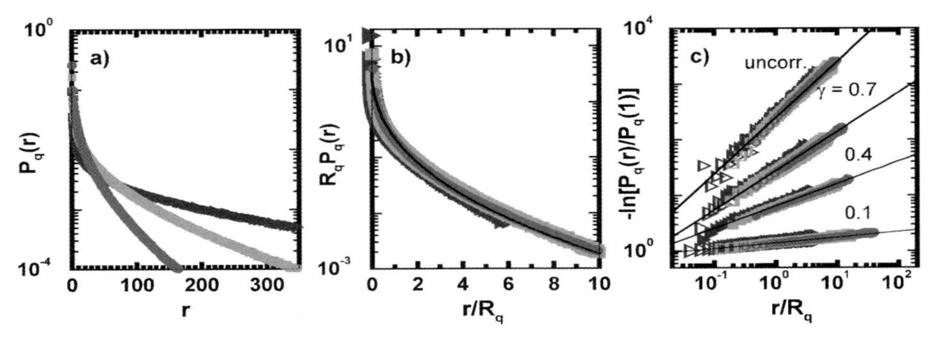

Fig. 4.3 (a) Distribution density $P_q(r)$ of the return intervals r for three thresholds $q = 1.5$ ($R_q \cong 15$, circle), 2.0 (44, square), and 2.5 (161, triangle) for long-term correlated data ($\gamma = 0.4$). (b) When plotting $P_q(r)$ multiplied by R_q versus r/R_q, the three curves collapse to a single stretched exponential curve (black line). (c) Double-logarithmic plot of $-\ln[Pq(r)/Pq(1)]$ as a function of r/R_q for $\gamma = 0.1, 0.4, 0.7$ as well as for the shuffled data (from bottom to top) and $q = 1.5$ (circle), 2.0 (square), and 2.5 (triangle). The straight lines are shown for comparison and have the slope γ for the long-term correlated data and 1 for the uncorrelated shuffled data.

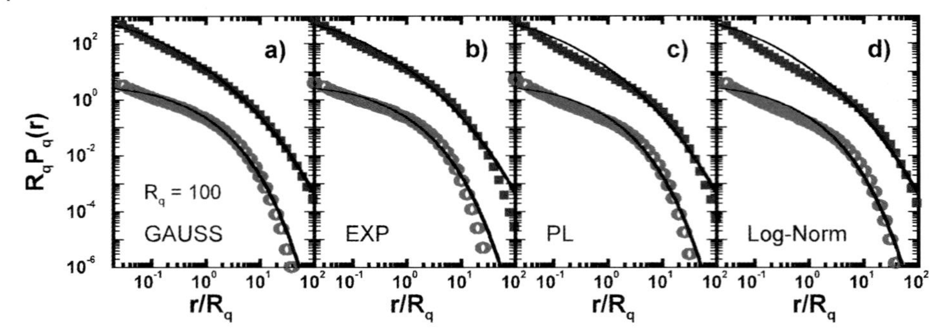

Fig. 4.4 Normalized rescaled distribution density functions $R_q P_q(r)$ of r-values with $R_q = 100$ as a function of r/R_q for long-term correlated data with $\gamma = 0.4$ (open symbols) and $\gamma = 0.2$ (filled symbols; we multiplied the data for the filled symbols by a factor 100 to avoid overlapping curves). In (a) the original data was Gaussian distributed, in (b) exponentially distributed, in (c) power-law distributed with power -5.5, and in (d) log-normally distributed. All four figures follow quite well stretched exponential curves (solid lines) over several decades. For small r/R_q values a power-law regime seems to dominate, while on large scales deviations from the stretched exponential behavior are due to finite-size effects. (Adapted from [31]).

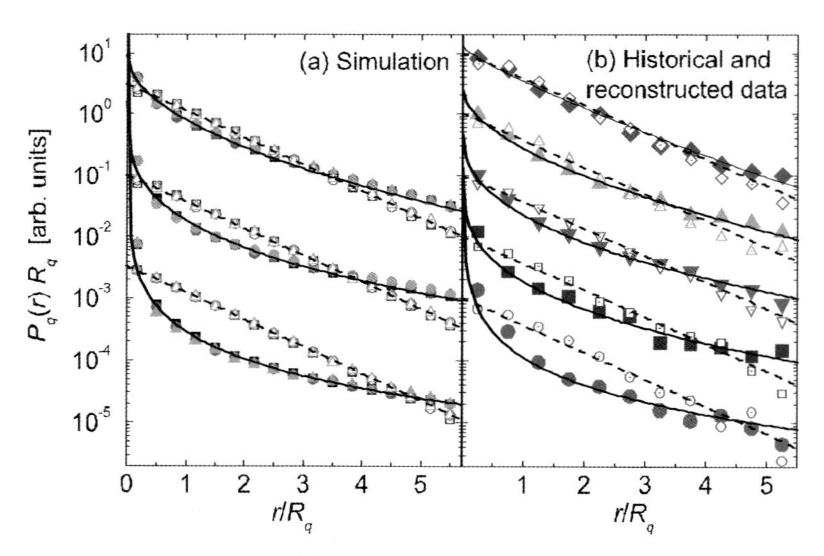

Fig. 4.5 (a) Distributions $P_q(r)$ of the return intervals r for the thresholds $q = 1.5$ ($R_q \approx 15$, squares), 2.0 ($R_q \approx 44$, circles), and 2.5 ($R_q \approx 161$, triangles) for simulated long-term correlated records with $\gamma = 0.5$ (top), 0.3 (middle), and 0.1 (bottom) (filled symbols) and for the corresponding shuffled data (open symbols). For the simulations, we used 1000 records of length $N = 2\,10^6$ for each value of γ. (b) Distribution densities $P_q(r)$ of the return intervals r for the five climate records considered in Fig. 4.1(g) with same symbols, for both original data (filled symbols) and shuffled data (open symbols). The data have been averaged over all quantiles q with $R_q > 3y$ and more than 50 return intervals. The lines are the theoretical curves following Eq. (4.3). (Adapted from [29]).

normal according to Eq. (4.2)) with correlation exponents $\gamma = 0.2$ and 0.4 in a double-logarithmic scale. The shape of the corresponding stretched exponentials, Eq. (4.2), is also plotted. The agreement is best for the Gaussian data, but also for the other distributions the stretched exponential fit is a good approximation over several decades.

Figure 4.5 compares $P_q(r)$ for simulated Gaussian data and three different correlation exponents with the five historical and reconstructed data sets introduced in Fig. 4.1. The solid lines, representing the theoretical curves with the measured γ values, match with $P_q(r)$ of the data (filled symbols). The dotted lines and the open symbols show the results for the shuffled data, when all correlations are destroyed. The shape of $P_q(r)$ becomes a simple exponential.

4.2.3
Power-Law Regime and Discretization Effects for Small Return Intervals

The curves in Fig. 4.4 exhibit significant deviations from the stretched exponential form also for small values of r ($r < R_q$) which we have studied in detail. It reveals the occurrence of an additional intermediate scaling regime for $r < R_q$. The scaling behavior with R_q still holds in this new regime, such that data for different values of q collapse onto each other as long as $r \gg 1$. The scaling behavior in this regime might be characterized by a power-law, giving rise to the two-branched ansatz

$$R_q P_q(r) \sim \begin{cases} (r/R_q)^{\gamma'} & \text{for} \quad 1 \ll r < R_q \\ \exp\left[-b_\gamma \left(r/R_q\right)^{\gamma}\right] & \text{for} \quad R_q < r \ll N, \end{cases} \tag{4.4}$$

replacing Eq. (4.2). For all distribution densities $P(x)$, $\gamma' \approx \gamma$ seems to be consistent with the data. However, we cannot fully exclude that γ' might depend slightly on the quantile q.

The behavior of $P_q(r)$ as described by Eq. (4.2) and the large r branch of Eq. (4.3) becomes visible only for very long data. However, for short natural records or large values of q and R_q, where the asymptotic regime $R_q < r \ll N$can hardly be observed because N does not significantly exceed R_q the power-law regime will clearly dominate the scaling behavior of the return interval distribution density. Hence, power-law distributions of return intervals, e.g., observed in a recent analysis of waiting times [43, 44] might be considered as a possible indication of long-term correlations.

For very small r values (close to 1) the continuous distribution density Eq. (4.3) has to be replaced by a discrete distribution. Hence, the data points for r close to 1 cannot obey the scaling of the distribution density with R_q, and no data collapse can be achieved for them. For the power-law distributed data and the log-normal distributed data, the discretization effects even seem to suppress the full development of the power-law scaling regime.

4.2.4
long-Term Correlations of the Return Intervals

The form of the distribution density $P_q(r)$ of return intervals between extreme events in long-term correlated data indicates that very short and very long return intervals r are more frequent than for uncorrelated data. However, $P_q(r)$ does not quantify, if the return intervals themselves are arranged in a correlated fashion, and if clustering of rare events may be induced by long-term correlations. In our previous work [29, 33] we reported that (1) long-term correlations in a Gaussian time series induce long-term correlations in the sequence of return intervals and (2) that both correlation functions are characterized by the same correlation exponent γ. We showed that this leads to a clustering of extreme events, an effect that also can be seen in long climate records.

In order to determine the autocorrelation behavior of the return series r_j, $j = 1, 2, \ldots, N_q$ for a given quantile q we calculate the autocorrelation function

$$C_r(s) = \frac{1}{\sigma_r^2 (N - s)} \sum_{j=1}^{N_q - s} (r_j - R_q)(r_{j+s} - R_q) \tag{4.5}$$

Figure 4.6 shows $C_r(s)$ and $C_x(s)$ for data characterized by the four distribution densities $D(x)$ and four return periods R_q. One can see that for each distribution, the data collapse to a single line, exhibiting the same slope as the original data. This shows that the return intervals are also long-term correlated, with the same value of γ as the original data. There is, however, one important difference in the correlation behavior: For the return intervals, the autocorrelation function $C_r(s)$ is significantly below the autocorrelation function $C_x(s)$ of the original data (also shown in Fig. 4.6) by a factor between 2 and 3, depending on the distribution. Accordingly, there is additional white noise in the return interval sequences that does not depend on the return period R_q. We believe that this uncorrelated component is a consequence of the way the return intervals are constructed. Tiny changes in the threshold q can lead to large changes of several return intervals in the records, and this causes the additional random component. The strength of this random component is independent of the threshold q as confirmed by the scaling of the autocorrelation function $C_r(s)$ for different R_q. There is no crossover in the scaling behavior of the autocorrelation function except for finite-size effects.

Alternatively to an autocorrelation analysis via autocorrelation function one might want to employ the detrended fluctuation analysis (DFA, see introduction) and use the relation $\alpha = 1 - \gamma/2$ between the fluctuation exponent α

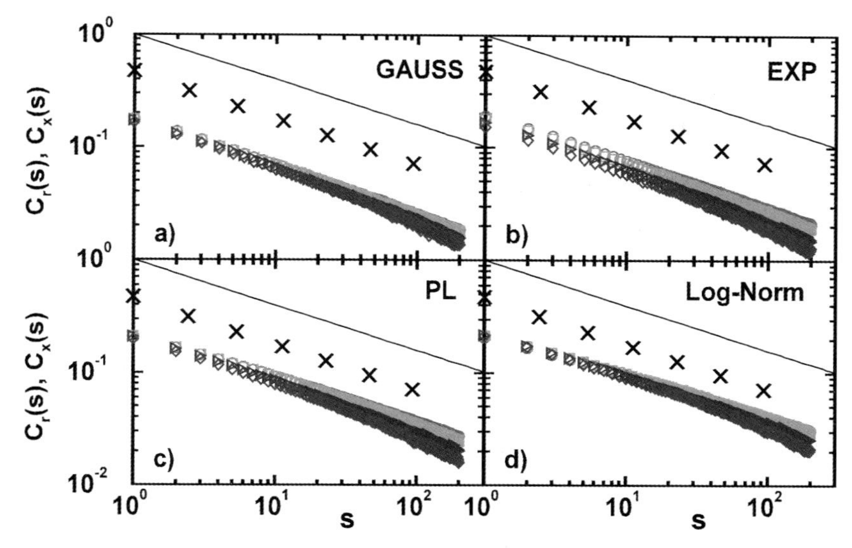

Fig. 4.6 Autocorrelation function $C_x(s)$ of the original data (cross) and $C_r(s)$ (see Eq. (4.5)) of series of return intervals: (a) Gaussian, (b) exponentially, (c) power-law, and (d) log-normally distributed original data with $\gamma = 0.4$ for $R_q = 10$ (circle), 20 (square), 50 (triangle), and 100 (diamonds). All curves show a power-law behavior indicating long-term correlations in the sequence of values with the same correlation exponent given in the original data. Curves for large R_q show weaker correlations due to finite-size effects. All results were averaged over 150 configurations of original data sets of length $N = 221$. Solid lines are guides to the eye with slopes $-\gamma = -0.4$. (Taken from [31]).

and the correlation exponent γ to determine γ [15, 18, 30]. In this case, however, a crossover is observed in the scaling behavior of the fluctuation function $F(s)$ because the uncorrelated component affects $F(s)$ also for $s > 1$. Hence, the classical autocorrelation analysis is more suitable for the study of the correlation properties of the return interval series than sophisticated DFA approaches.

4.2.5
Conditional Return Intervals

The long-term correlations in the sequence of the return interval series studied in the previous Section cause a dependence of the probability of finding a certain return interval r on the history. In order to study this effect we considered the *conditional* distribution density $P_q(r|r_0)$, that is the distribution density of all those return intervals that directly follow a return interval of given size r_0. An explanation of conditional return intervals is shown in Fig. 4.7. For fixed r_0/R_q the data points still collapse to single curves, independent of the distribution of the data. Taking Eq. (4.3) into account our results show that the conditional distribution density function scales as

$$P_q(r|r_0) \cong \frac{1}{R_q} f_{r_0/R_q}\left(r/R_q\right) . \tag{4.6}$$

The long-term correlations in the sequence of r values cause a culmination of small $r$values for small r_0 and large r values for large r_0. The conditional distribution densities for the five historical and reconstructed records and for artificial data with the same correlation exponents and similar set lengths are show in Fig. 4.8 for just two conditions, r_0 smaller $(-)$ or larger $(+)$ the median r value. The splitting of the curves $P_q^+(r)$ and $P_q^-(r)$ for large r/R_q is a consequence of the long-term correlations.

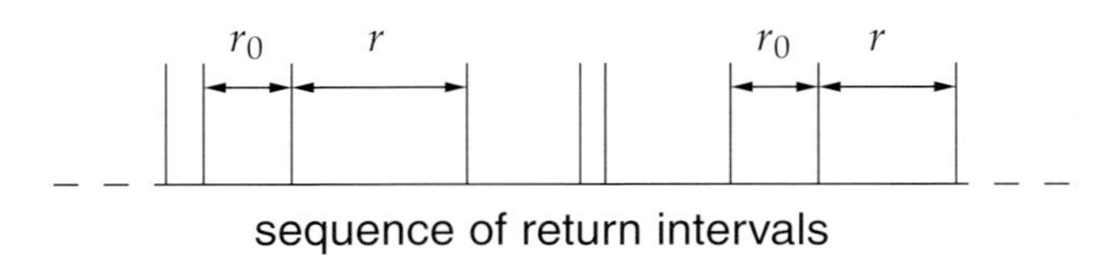

Fig. 4.7 Conditional return intervals are those r values, that follow in the sequence of the return intervals directly after a r value of a distinct size r_0. For data with poor statistics, it is convenient to use r_0 as a boundary value, i.e., considering all those r values that follow directly after a r value smaller or larger than, e.g., the median of all return intervals.

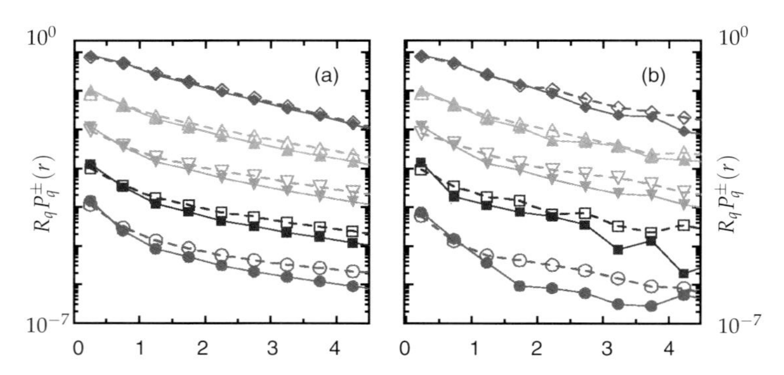

Fig. 4.8 Conditional distribution densities $R_q P_q^+(r)$ (open symbols) and $R_q P_q^-(r)$ (full symbols), averaged over all r_0 above and below the median return interval, respectively, versus r/R_q for (a) artificial records and (b) the five climate records from Fig. 4.1. The artificial data in (b) have the same γ values and mean record lengths as the climate records; we studied 1000 records of size $N = 1250$. (Adapted from [29]).

4.2.6
Conditional Mean Return Intervals

In contrast to R_q the conditional mean return interval $R_q(r_0) = \sum_{r=1}^{\infty} r P_q(r|r_0)$, i.e. the average return interval of those r values that follow directly after an interval of size r_0, clearly exhibits correlation effects. Figure 4.9 shows $R_q(r_0)$ in units of R_q as a function of r_0/R_q for four values of R_q for long-term correlated data ($\gamma = 0.4$) following the four distribution densities $P(x)$ listed in Eq. (4.2). The correlation effect becomes apparent: after small r_0/R_q the next expected return interval $R_q(r_0)$ is smaller than R_q, and after large r_0/R_q it is much larger than R_q. Although the shapes of the (more or less) collapsing curves differ depending on the original distribution and on the chosen quantile q (i.e. on R_q), the tendency is the same for all four original distribution densities.

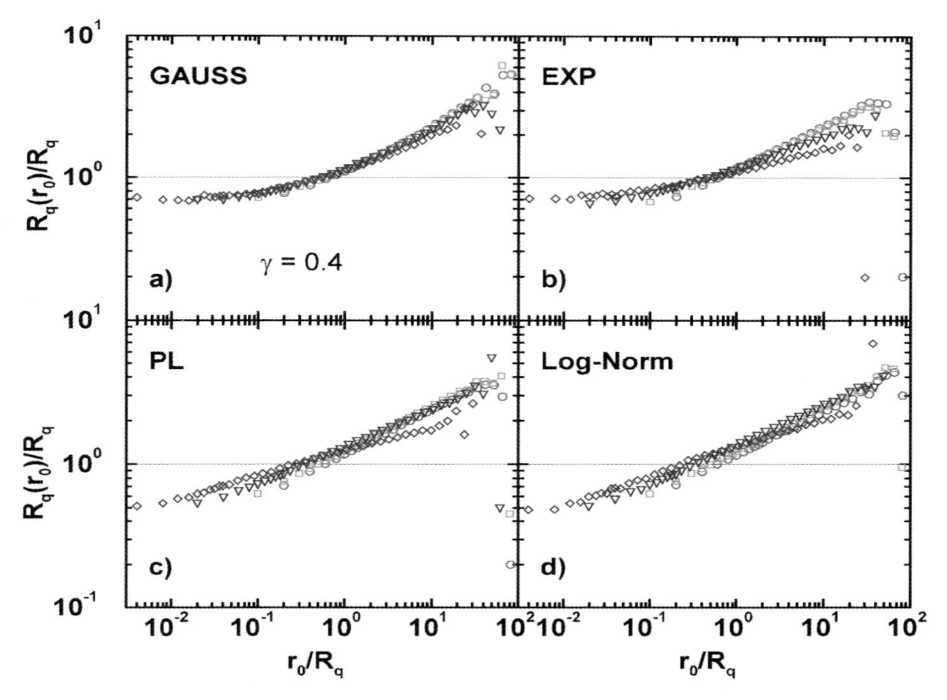

Fig. 4.9 Conditional return periods $R_q(r_0)$ in units of $R_q$versus the condition r_0/R_q for $R_q = 5$ (circles), 10 (squares), 50 (triangles), and 250 (diamonds) for $\gamma = 0.4$. While Gaussian (a) and exponentially (b) distributed data show a data collapse with nearly no R_q-dependence (except for stronger finite-size effects for exponential data at large r_0/R_q), power-law (c) and log-normally distributed (d) data show slightly different slopes for different R_q. All figures display the memory-effect in form of increasing $R_q(r_0)$ with increasing r_0. In uncorrelated data $R_q(r_0) = R_q$, as indicated by the horizontal lines at ratio one. (Taken from [31]).

By definition, $R_q(r_0)$ is the expected waiting time to the next event, when the two events before were separated by r_0. A more general quantity is the expected residual waiting time $\tau_q(t|r_0)$ to the next event, when the time t has been elapsed (see Fig. 4.10). For $t = 0$, $\tau_q(t|r_0)$ is identical to $R_q(r_0)$. In general, $\tau_q(t|r_0)$ is related to $P_q(r|r_0)$ by

$$\tau_q(t|r_0) = \int_x^\infty (r - t) P_q(r|r_0)\, dr \Big/ \int_x^\infty P_q(r|r_0)\, dr. \tag{4.7}$$

For uncorrelated records, $\tau_q(t|r_0) = R_q$ (except for discreteness effects that lead to $\tau_q(t|r_0)/R_q > 1$ for $t > 0$, see [45]. Due to the scaling of $P_q(r|r_0)$, we expect that also $\tau_q(t|r_0)/R_q$ scales with r_0/R_q and t/R_q.

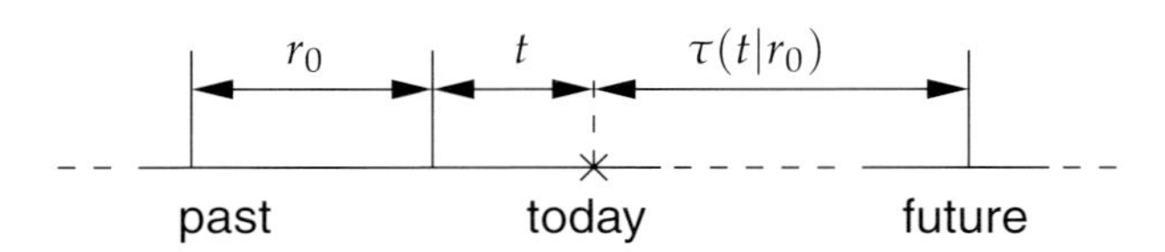

Fig. 4.10 The conditional residual waiting time $\tau_q(t|r_0)$ is defined as the average remaining time between today and the next extreme event, when the time between today and the last extreme event is x and the return interval between the two last extreme events is r_0.

Figure 4.11(a) shows that this is indeed the case. The data collapse for each value of t/R_q confirms the scaling property. The figure clearly displays the effect of the long-term memory: Small and large return intervals are more likely to be followed by small and large ones, respectively, and hence $\tau_q(t|r_0)/R_q \equiv R_q(r_0)/R_q$ is well below (above) 1 for r_0/R_q well below (above) 1. With increasing t, the expected residual time to the next event increases, as is also shown in Fig. 4.11(b), for two values of r_0 (top and bottom curve). Note that only for an infinite long-term correlated record, the value of $\tau_q(t|r_0)$ will increase indefinitely with t and r_0. For real (finite) records, there exists a maximum return interval which limits the values of t, $r_0$and $\tau_q(t|r_0)$. The middle curve shows the expected residual time averaged over all r_0, i.e. the unconditional residual time. In this case, the interval between the last two events is not taken explicitly into account, and the slower-than-Poisson-decrease of the unconditional distribution density $P_q(r)$, leads to the anomalous increase of the mean residual waiting time with the elapsed time [45]. Very recently, this approach (average over r_0) has been applied successfully to worldwide earth quake records [36] (see Section 4). For the case of long-term correlated records, however, like the hydroclimate records discussed here, the large differences between the three curves in Fig. 4.11(b) suggest that for an efficient

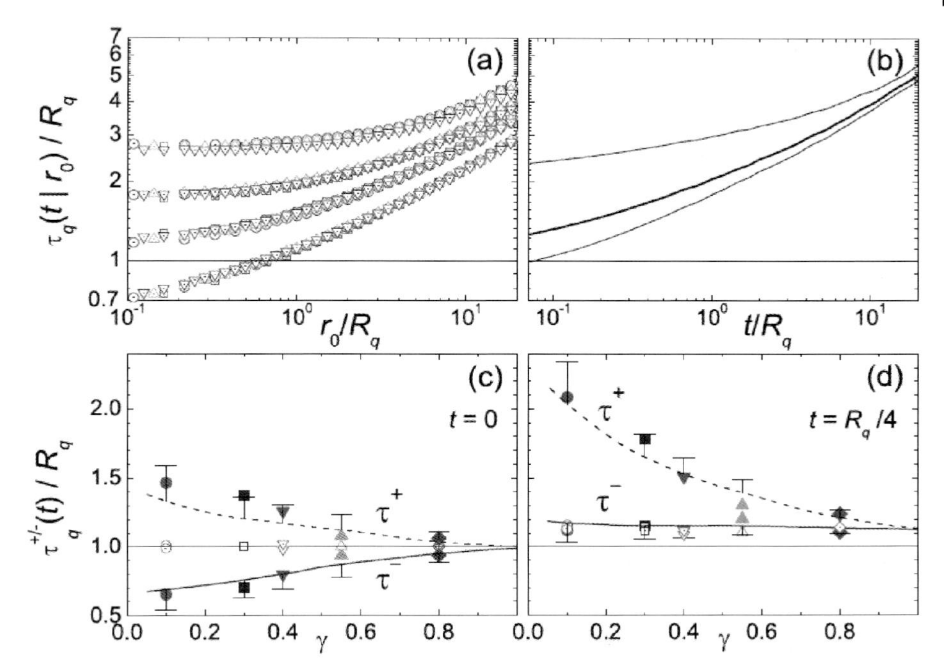

Fig. 4.11 (a) Mean residual time to the next event $\tau_q(t|r_0)$ (in units of R_q) versus r_0/R_q for four q values ($q = 1.0, 1.25, 1.5$ and 1.75, different symbols) and four values of the elapsed time x since the last event ($t/R_q = 0, 0.25, 1$ and 4, from bottom to top). (b) Mean residual time $\tau_q(t|r_0)$ as a function of t/R_q, for $r_0/R_q = 1/8$ (lower curve) and $r_0/R_q = 8$ (upper curve). The middle curve represents the mean residual time averaged over all r_0. (c, d) Mean residual times $\tau_q^-(x)$ (full line) and $\tau_q^+(x)$ (dashed line), averaged over all r_0 below and above the median return interval, respectively, for (c) $t = 0$ and (d) $t = R_q/4$. The symbols are for the five climate records from Fig. 4.1, for both original data (filled symbols) and shuffled data (open symbols). (Adapted from [29]).

risk estimation, also the previous return interval has to be taken into account, and not only the distribution of the return intervals.

To reveal this intriguing behavior in the relatively short observed and reconstructed records, we improved the statistics (similar to Fig. 4.8) by studying the mean residual waiting times $\tau_q^-(t)$ and $\tau_q^+(t)$ for r_0 below and above the median, respectively. For uncorrelated data, both quantities are identical and coincide with R_q.

Figure 4.11(c) shows $\tau_q^-(0)/R_q$ and $\tau_q^+(0)/R_q$ versus γ for simulated records (lines) and the five representative climate records (symbols). The difference between $\tau_q^-(t)$ and $\tau_q^+(t)$ becomes more pronounced with decreasing value of γ, i.e. increasing long-term memory. The results for the climate records are in good agreement with the theoretical curves. The same comparison for $t/R_q = 1/4$ instead of $t = 0$ is shown in Fig. 4.11(d). The behavior is qualitatively different: while $\tau_q^-(0)/R_q$ increases with increasing

γ, $\tau_q^-(R_q/4)/R_q$ is rather constant. Again, the agreement between simulated and real records is quite satisfactory, revealing the strong effect of memory in the hydroclimate records that also results in the clustering of the extreme events. To show the significance of the results, we also analyzed the corresponding shuffled data. We obtained $\tau_q^+(0)/R_q \approx \tau_q^-(0)/R_q \approx 1$ and $\tau_q^+(R_q/4)/R_q \approx \tau_q^-(R_q/4)/R_q \approx 1.1$. In the second case, the shuffled data (following the Poisson distribution) show a slight increase of the residual time (1.1 instead of 1). This is a finite-size effect that has already been noticed in [45].

4.3 Statistics of Maxima

Extreme events are rare occurrences of extraordinary nature, such as floods, very high temperatures, or major earthquakes. In studying the extreme value statistics of the corresponding time series one wants to learn about the distribution of the extreme events, i.e., the maximum values of the signal within time segments of fixed duration R, and the statistical properties of their sequences. Figure 4.12 illustrates the definition of the series of maxima (m_j), $j = 1, \ldots, N/R$ of original data (x_i), $i = 1, \ldots, N$, within segments of size R for $R = 365$, i.e. for annual maxima if (x_i) represents daily data.

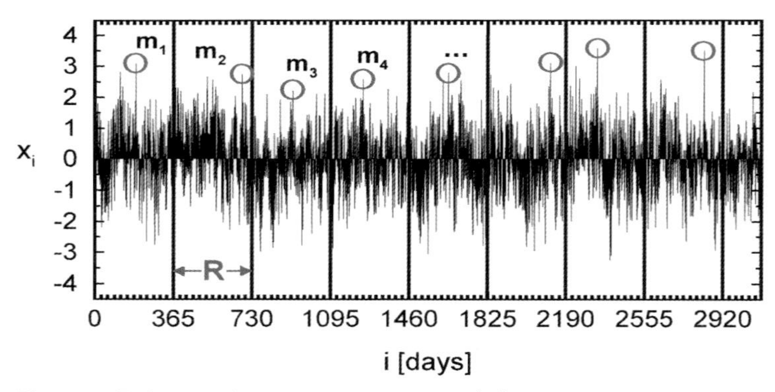

Fig. 4.12 Definition of maxima: A time series (x_i), $i = 1, \ldots, N$, of, e.g., daily data is separated into segments of length $R = 365$ days. The maximum values m_j (circles) in each segment, e.g., annual maxima, define another time series (m_j), $j = 1, \ldots, N/R$. (Taken from [34]).

Many exact and empirical results on extreme value statistics have been obtained in the past years, for reviews see, e.g., [42,46–50]. Most of these results, however, hold only in the limit $R \to \infty$ and are based on statistically independent values of the time series. Both assumptions are not strictly valid in practice. Since observational data are always finite, predictions for finite time intervals R are required, and – most importantly – correlations cannot be disregarded [34].

4.3.1
Extreme Value Statistics for i.i.d. Data

In classical extreme value statistics one assumes that records (x_i) consist of independently and identically distributed (i.i.d.) data, described by density distributions $P(x)$, which can be, e.g., a Gaussian or an exponential distribution. One is interested in the distribution density function $P_R(m)$ of the maxima (m_j) determined in segments of length R in the original series (x_i) (see Fig. 4.12). Note that all maxima are also elements of the original data. The corresponding integrated maxima distribution $G_R(m)$ is defined as

$$G_R(m) = 1 - E_R(m) = \int_{-\infty}^{m} P_R(m')\, dm. \tag{4.8}$$

Since $G_R(m)$ is the probability of finding a maximum smaller than m, $E_R(m)$ denotes the probability of finding a maximum that exceeds m. One of the main results of traditional extreme value statistics states that the integrated distribution $G_R(m)$ converges to a double exponential Fisher–Tippet–Gumbel distribution (often labeled as Type I) [46–48, 51] for i.i.d. data with Gaussian or exponential $P(x)$, i.e.,

$$G_R(m) \to G\left(\frac{m-u}{\alpha}\right) = \exp\left[-\exp\left(-\frac{m-u}{\alpha}\right)\right] \tag{4.9}$$

for $R \to \infty$, where α is the scale parameter and u the location parameter. By the method of moments the parameters are given by $\alpha = \sigma_R\sqrt{6}\big/\pi$ and $u = m_R - n_e\alpha$ with the Euler constant $n_e = 0.577216$ [48, 52–54].

Here m_R and σ_R denote the (R dependent) mean maximum and the standard deviation, respectively. Note that different asymptotics will be reached for broader distributions of data (x_i) that belong to other domains of attraction [48]. For example, for data following a power-law or Pareto distribution, $P(x) = (x/x_0)^{-k}$, $G_R(m)$ converges to a Fréchet distribution, often labelled as Type II. For data following a distribution with finite upper endpoint, for example the uniform distribution $P(x) = 1$ for $0 \leq x \leq 1$, $G_R(m)$ converges to a Weibull distribution, often labelled as Type III. These are the other two types of asymptotics, which, however, we do not consider here.

4.3.2 Effect of Long-Term Persistence on the Distribution of the Maxima

In long-term correlated records extreme events cannot be viewed a priori as uncorrelated even when there is a long time span between them. Recently, there have been some approaches to include correlations in the study of extreme value statistics. For the special case of Gaussian $1/f$ correlations in voltage fluctuations in GaAs films extreme value statistics have been demonstrated to follow a Gumbel distribution [55]. A somewhat different asymptotic behavior was observed in experiments on turbulence and in the two-dimensional XY model [56, 57], see also [58]. Extreme value statistics have also been employed in studies of hierarchically correlated random variables representing the energies of directed polymers [59] and in studies of maximal heights of growing self-affine surfaces [60]. In the Edwards–Wilkinson model and the Kardar–Parisi–Zhang model for fluctuating, strongly correlated interfaces an Airy distribution function has been obtained as exact solution for the distribution of maximal heights very recently [61, 62]. On the other hand, the statistics of extreme height fluctuations for Edwards–Wilkinson relaxation on small-world substrates are rather described by the classical Fisher–Tippet–Gumbel distribution [63]. Besides these recent results there is a theorem by S. M. Berman [64] (see also [47, 48]) stating that the maxima statistics of stationary Gaussian sequences with correlations converges to a Gumbel distribution asymptotically for $R \to \infty$ provided that $C_x(s)\log(s) \to 0$ for $s \to \infty$, which holds for long-term correlations.

Figure 4.13 shows how the convergence of the integrated maxima distribution $G_R(m)$ towards the Gumbel distribution, Eq. (4.7), is affected by long-term correlations in exponentially distributed signals (x_i). In panels (a, b) the unscaled distribution densities $P_R(m)$ of the maxima within segments of size R are shown for several values of R. Since Eqs. (4.6) and (4.7) yield that for $R \to \infty$

$$P_R(m) \to \frac{1}{\alpha} \exp\left[-\exp\left(-\frac{m-u}{\alpha}\right) - \frac{m-u}{\alpha}\right], \tag{4.10}$$

the distribution densities $P_R(m)$ can be scaled upon each other if $\alpha P_R(m)$ is plotted versus $(m-u)/\alpha$, see Fig. 4.13(c, d). In Fig. 4.13(c, d) deviations occur only at very small R values ($R < 10$) where scaling breaks down due to the sharp cut-off of the exponential density distribution $P(x)$ at $x = 0$. In the long-term correlated case, where the correlation time T diverges, the fast convergence is particularly surprising, since the segment duration R can never exceed T. From a theoretical point of view, we expect a convergence towards the Gumbel limit only for very large R values. The reason for this fast convergence may be a rapid weakening of the correlations among the maxima with increasing values of R.

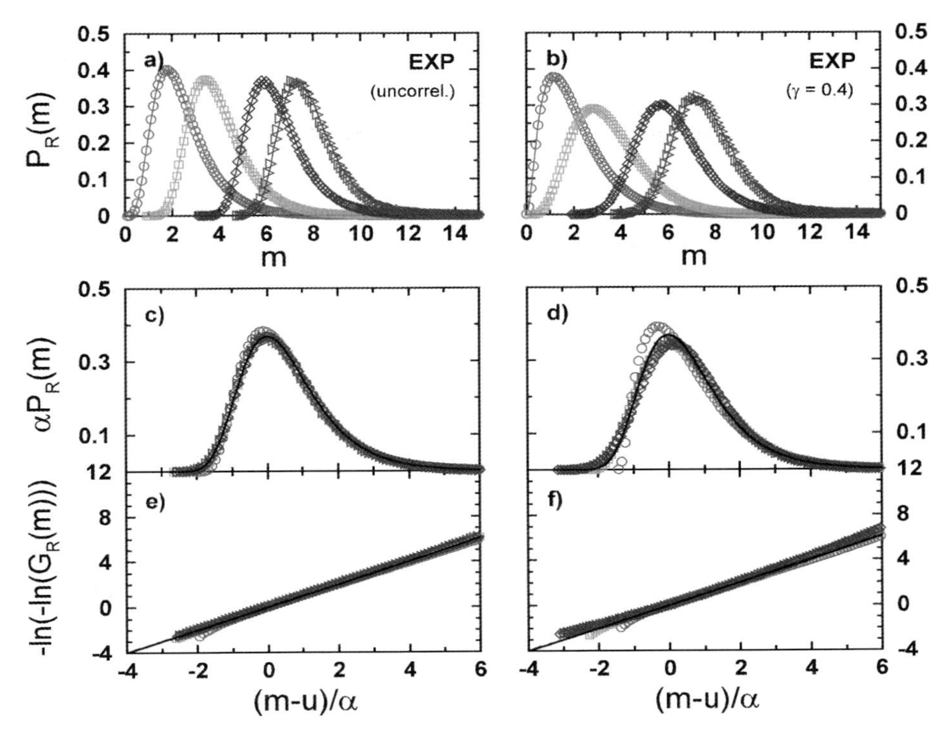

Fig. 4.13 Distributions of maxima in segments of length R for (a, c, e) uncorrelated and (b, d, f) long-term correlated ($\gamma = 0.4$) data with exponential distribution. Panels (a, b) show the distribution density function PR(m) of the maximum values for four segment sizes $R = 6$ (circles), 30 (squares), 365 (diamonds), and 1500 (triangles). Panels (c, d) show that a collapse of all four curves to a single curve is achieved in both cases, when the m axis is replaced by $(m-u)/\alpha$ and $P_R(m)$ is multiplied by the scale parameter α. The solid line is the Gumbel distribution density, Eq. (4.10). Panels (e, f) show the corresponding integrated distribution $G_R(m)$ together with the Gumbel function Eq. (4.9). (Adapted from [34]).

In Fig. 4.13(e, f) it is shown that the convergence towards Eq. (4.7) (continuous line) is also fast for the exponentially distributed data. In contrast, the convergence is much slower in the case of a Gaussian distribution of the original data [34]. Hence, the distribution $P(x)$ of the original data has a much stronger effect upon the convergence towards the Gumbel distribution than the long-term correlations in the data.

4.3.3
Effect of Long-Term Persistence on the Correlations of the Maxima

The distributions of maxima considered in the previous Section do not quantify, however, if the maxima values are arranged in a correlated or in an uncorrelated fashion, and if clustering of maxima may be induced by long-term correlations in the data. To study this question, we have evaluated the correlation properties of the series of maxima (m_j), $j = 1, \ldots, N/R$, of long-term correlated data with Gaussian distribution. Figure 4.14(a) shows representative results for the maxima autocorrelation function

$$C_m(s) = \frac{\langle (m_j - m_R)(m_{j+s} - m_R) \rangle}{\langle (m_j - m_R)^2 \rangle}, \tag{4.11}$$

where m_R denotes the average maximum value in the series, and $\langle \ldots \rangle$ is the average over j similar to Eq. (4.1). The comparison with the scaling behavior of the auto-correlation function $C_x(s)$ of the original data (x_i) (see Eq. (4.1)) that follows a power law decay, $C_x(s) \sim s^{-\gamma}$ with $\gamma = 0.4$, reveals the presence of long-term correlations with a correlation exponent $\gamma' \approx \gamma$ in the maxima series. Hence, large maxima m are more likely to be followed by large maxima and small maxima are rather followed by small maxima, leading to clusters of large and small maxima.

Figures 4.14(b, c) show that the deviations of the auto-correlation function $C_m(s)$ from a power-law fit with slope $\gamma = -0.4$ for large values of R and s are presumably caused by finite-size effects. They become significantly smaller as the length N of the series is increased. In the case of uncorrelated data, the series of maxima is also uncorrelated, $C_m(s) = 0$ for $s > 0$ (not shown).

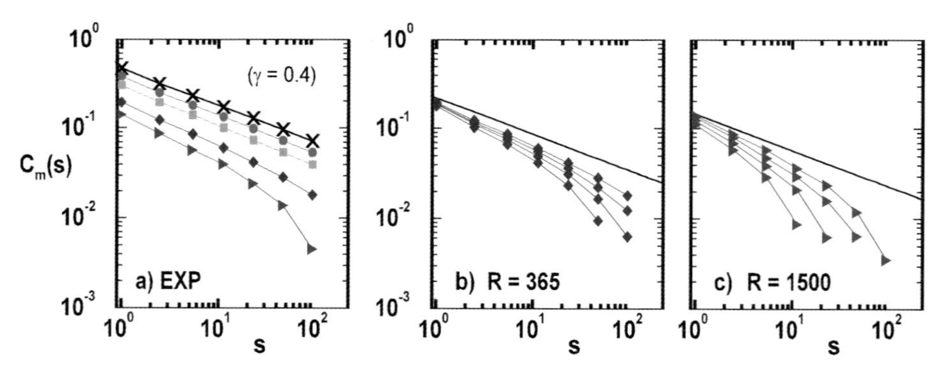

Fig. 4.14 (a) Auto-correlation function $C_m(s)$ of the maxima (m_j) of Gaussian distributed (x_i) for different R values, $R = 6$ (circles), 30 (squares), 365 (diamonds), and 1500 (triangles). The auto-correlation function $C_x(s)$ of the original data (x_i) (crosses) shows the slope $-\gamma = -0.4$. (b, c) Study of finite-size effects with (b) $R = 365$ and (c) 1500. The set lengths are $N = 221$ (circles), 220 (squares), 219 (diamonds), and 218 (triangles). The descent of the slopes of $C_m(s)$ from the slope of the straight line ($-\gamma = -0.4$) with decreasing set length seems to be a finite-size effect. (Adapted from [34]).

4.3.4 Conditional Mean Maxima

As a consequence of the long-term correlations in the series of maxima (m_j), the probability of finding a certain value m_j depends on the history, and in particular on the value of the immediately preceding maximum m_{j-1}, which we will denote by m_0 in the following. This effect has to be taken into account in predictions and risk estimations. For a quantitative analysis we considered conditional maxima as illustrated in Fig. 4.15, where all maxima following an $m_0 \approx 6$ (within the grey band), i.e., the subset of maxima which fulfill the condition of having a preceding maximum close to m_0, are indicated by circles. The width Δm_0 sketched by the grey band around m_0 in Fig. 4.15 is set such that a sufficient number of approximately 700 conditional maxima is obtained for each record. The corresponding conditional mean maximum value $m_R(m_0)$ is defined as the average of all these conditional maxima. Note that $m_R(m_0)$ will be independent of m_0 for uncorrelated data.

Figure 4.16 shows the conditional mean maxima $m_R(m_0)$ versus m_0 for long-term correlated Gaussian and exponentially distributed data for four values of R. Of course, the mean maxima are larger for larger segment sizes R. This dependence is also observed for the unconditional mean maxima indicated by horizontal lines in Fig. 4.15. In addition to this trivial dependence, the conditional mean maxima significantly depend upon the condition, i.e. the previous maximum m_0, showing a clear memory effect. Evidently, this dependence is most pronounced for the small segment durations. However, it is still observable for the large $R = 365$ (most common for observational daily data) and even $R = 1500$ (beyond common observational limits). Note that the results for Gaussian and exponentially distributed data agree only qualitatively:

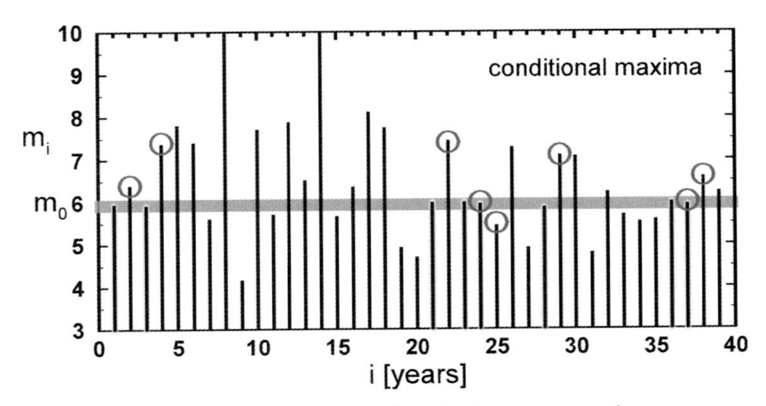

Fig. 4.15 Definition of conditional maxima. In the sequence of (annual) maxima only those m values (indicated by circles) are considered, which directly follow a maximum of approximate size $m_0 \approx 6$ (gray band). The new sequence of m values is the sequence of conditional maxima. (Taken from [34]).

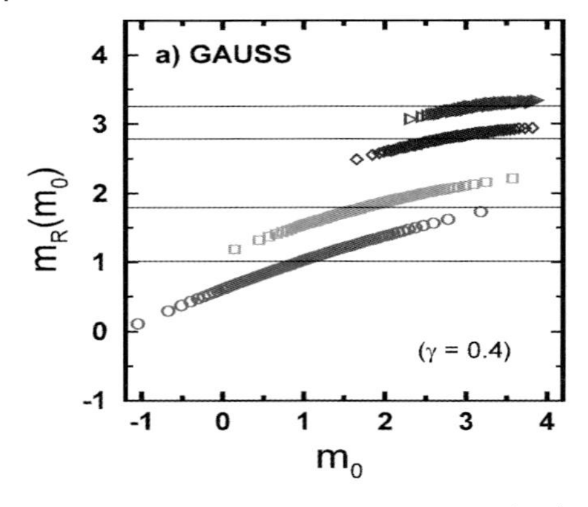

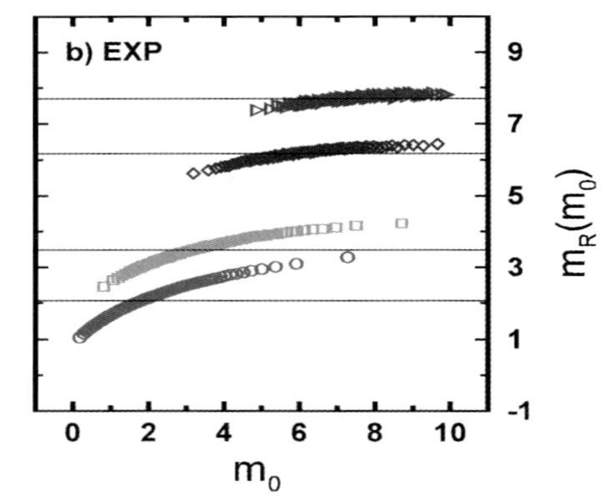

Fig. 4.16 Mean conditional maxima $m_R(m_0)$ for $\gamma = 0.4$ and $R = 6$ (circles), 30 (boxes), 365 (diamonds), and 1500 (triangles) versus m_0 for(a) Gaussian and (b) exponential data. The straight lines represent the unconditional means m_R for given R. The width Δm_0 for the condition m_0 was chosen such that approximately 700 m values were obtained for each m_0 in each of the 150 runs of $N = 2^{21}x$ values. Both figures show the memory effect in form of $m_R(m_0) > m_R$ for rather large m_0 (above m_R) and $m_R(m_0) < m_R$ for rather small m_0 (below m_R). (Taken from [34]).

while the $m_0$dependence of $m_R(m_0)$ is quite close to linear for Gaussian data, there seems to be significant curvature for the exponentially distributed data, which is a remnant of the asymmetry of the exponential distribution.

To test our predictions on real records with long-term correlations we studied the annual data of the Nile river water level minima ([25]; see also Fig. 4.1) and the reconstructed northern hemisphere annual temperatures by Moberg [65]. Since the Nile data consists of annual minima, we studied extreme minima instead of maxima. Both records are long-term correlated with $\gamma \approx 0.3$ in Eq. (4.1).

In order to get sufficient statistics for the conditional means $m_R(m_0)$, we have considered six m_0 intervals for each value of R and have set the width Δm_0 of the band around m_0 such that there are no gaps between the bands. Figure 4.17 shows the results for three values of R. In all cases, the effect of the long-term correlations (persistence) on the conditional mean minima and maxima $m_R(m_0)$is clearly visible for both records: the conditional means are smaller for smaller condition value m_0 and larger for larger condition value. As shown by the open symbols in Fig. 4.17 the m_0 dependence completely disappears for randomly shuffled surrogate data, indicating that the dependence was due to the correlations in the data.

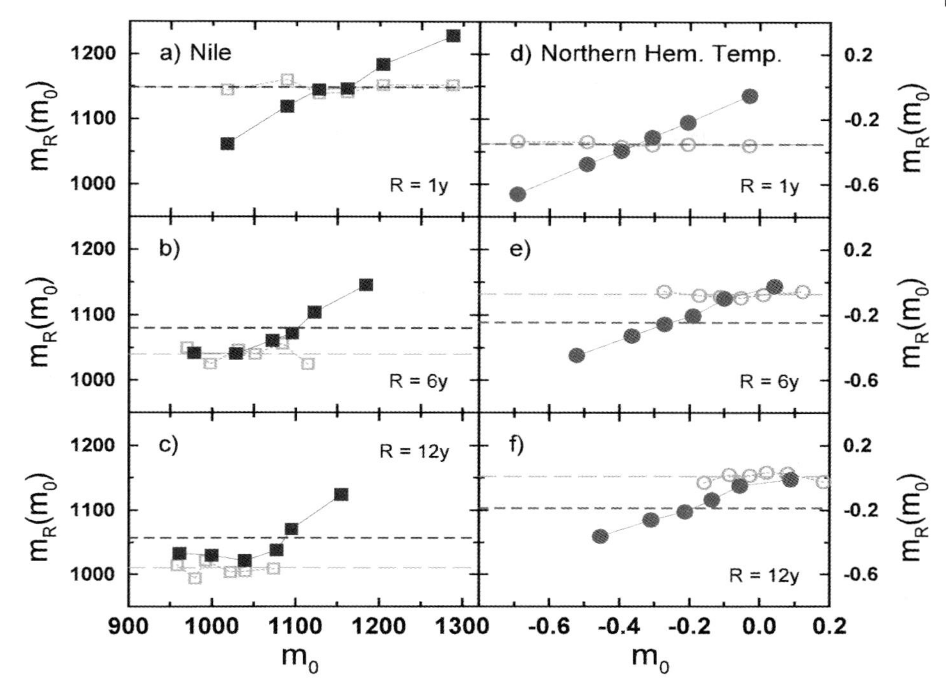

Fig. 4.17 (a–c) Mean conditional minima $m_R(m_0)$ for annual data of Nile river water level minima (squares) [21] and (d–f) mean conditional maxima for the reconstructed northern hemisphere annual temperatures after Moberg (circles) [64], for (a, d) $R = 1y$, (b, e) $6y$, and (c, f) $12y$. The filled symbols show the results for the real data and the open symbols correspond to surrogate data where all correlations have been destroyed by random shuffling. The unconditional mean minima (a–c) and maxima (d–f) are indicated by dashed lines; the long-dashed lines correspond to the shuffled data. (Taken from [34]).

4.3.5 Conditional Maxima Distributions

The quantity $m_R(m_0)$ is the first moment of the conditional distribution density $P_R(m|m_0)$, which is defined as the distribution density of all maxima m_j that follow a given maximum value m_0. Figure 4.18 shows $P_R(m|m_0)$ for two values of $m_0$and again for Gaussian as well as for exponentially distributed long-term correlated data sets with $\gamma = 0.4$. When compared with the unconditional distribution density $P_R(m)$, the long-term correlations lead to a shift of $P_R(m|m_0)$ to smaller m values for small $m_0$and to larger m values for large m_0, respectively. The conditional exceedance probability

$$E_R(m|m_0) = \int_m^\infty P_R(m'|m_0)\, dm' \qquad (4.12)$$

defines the probability of finding a maximum larger than m provided that the previous value was close to m_0. We found a strong dependence of $E_R(m|m_0)$

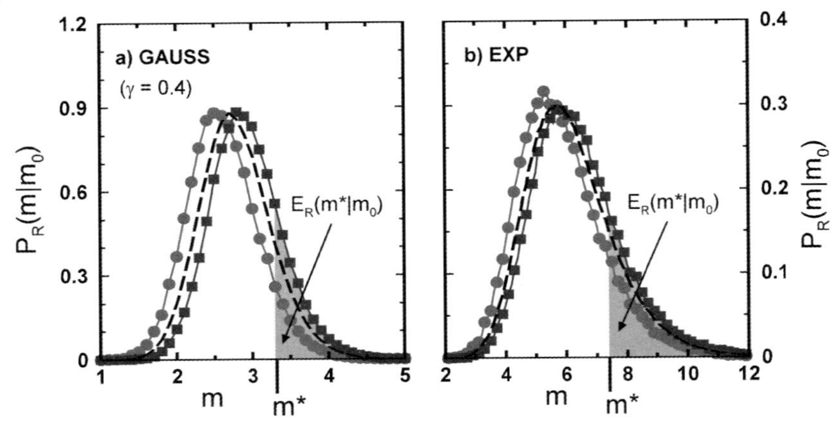

Fig. 4.18 (a) Conditional distribution density $P_R(m|m_0)$ of maximum values taken from correlated Gaussian data ($\gamma = 0.4$) with $R = 365$ and $m_0 = 2.06$ (circles) as well as $m_0 = 3.55$ (squares). (b) shows the same as (a) for exponentially distributed data with $m_0 = 4.10$ (circles) and $m_0 = 8.65$ (squares). The probability $E_R(m^*|m_0)$ to find a m value larger than an arbitrarily given m^* (see Eq. (4.12)) also depends on m_0. (Taken from [34]).

upon the conditionm_0. Consequently, the difference between the unconditional probabilities $E_R(m)$ and the corresponding conditional probabilities $E_R(m|m_0)$ depends strongly on m_0 in the presence of long-term correlations.

Figure 4.19 shows the ratios of the conditional exceedance probabilities $E_R(m|m_0)$ and the unconditional exceedance probabilities $E_R(m)$. The figure clearly shows an in crease of the memory effect for larger m values, i.e., for more extreme events. This increase seems weaker for exponentially distributed data than for Gaussian distributed data due to the less correlated maximum series of exponential data; however the tendency is the same. As Fig. 4.19 shows $E_R(m|m_0)$ can differ up to a factor of two from $E_R(m)$ when considering the history m_0 in the presence of long-term correlations (with $\gamma = 0.4$). This effect has to be taken into account in predictions and risk estimations of large events.

4.4 Long-Term Memory in Earthquakes

Understanding the mechanism behind the complex spatio-temporal behavior of earthquakes is one of the major challenges in science (for reviews, see, e.g., [37, 66–68]). So far, several empirical scaling laws have been established, including the Omori law [69], the Gutenberg–Richter law [70] and Bath's law [71]. The Omori law describes how the rate of aftershocks decays with

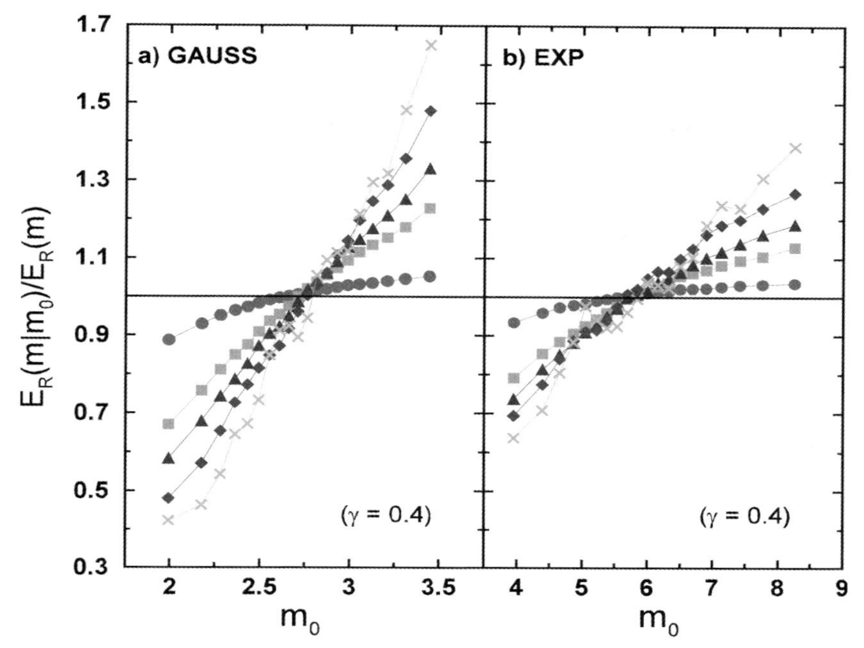

Fig. 4.19 Ratios of the conditional and unconditional exceedance probabilities $E_R(m|m_0)$ and $E_R(m)$ to find a maximum larger than m for (a) Gaussian and (b) exponentially distributed data with $\gamma = 0.4$ and $R = 365$. The six m values were chosen such that $E_R(m) = 0.9$ (circles, $m = 2.15$ for Gaussian and 4.40 for exponential data), 0.5 (squares, $m = 2.75$ and 5.95), 0.3 (triangles up, $m = 2.95$ and 6.70), 0.1 (diamonds, $m = 3.35$ and 8.00), and 0.01 (crosses, $m = 3.95$ and 10.40). Each point in the graph is based on a statistics of 500 conditional m values and averaged over 150 runs of $N = 2^{21}x$ values. The effect of long-term correlations seems to be strongest for the largest m (crosses): depending on $m_0 E_R(m|m_0)$ varies from 0.4 up to 1.7 for Gaussian data, i.e., by a factor greater than four. For exponential data this factor is still greater than two. (Taken from [34]).

time as a power law, and the Gutenberg–Richter law characterizes the (exponential) distribution of the earthquake magnitudes. Bath's law, finally, characterizes the typical difference in the magnitude between the mainshock and its largest aftershock.

A well-known phenomenological model that describes several aspects of seismic activity is the epidemic type aftershock sequence (ETAS) model of triggered seismicity [72–74]. In the ETAS model there is no distinction between mainshocks and aftershocks. Each earthquake event can trigger an aftershock sequence, and the length of the sequence as well as the distribution of the magnitudes and the rates of activity in this sequence are governed by the three laws mentioned above.

Recently, scaling laws for the temporal and spatial variability of earthquakes in several seismic regions with different tectonic properties have been obtained by Bak et al. [75] and Corral [36, 76, 77]. Considering the various tectonic environments as well as mainshocks and aftershocks as part of essentially one unique process, they analyzed the return intervals between earthquakes with amplitudes greater than M in a large number of spatial areas of varying sizes $L \times L$. While Bak et al. [75] concentrated on the distribution of the return intervals in California and obtained a unified scaling law for the spatio-temporal set of data (see also [78, 79]), Corral [36, 76, 77] studied the return intervals in a large number of spatial areas of various sizes with stationary seismic activity. He found that independent of the considered area and independent of the threshold M, the distribution $D_M(r)$ of return intervals scales with the mean inter-occurrence time R_M as

$$D_M(r) = \frac{1}{R_M} f\left(r/R_M\right), \tag{4.13}$$

where $f(x)$ is a universal scaling function which does not depend on M. Corral found that (apart from very small rescaled return intervals $x = r/R_M$), $f(x)$ can be well approximated by the Gamma distribution

$$f(x) = Cv^{-(1-\gamma)} \exp\left(-x^{\delta}/B\right), \tag{4.14}$$

with $C \cong 0.5$, $B \cong 1.58$, $\delta \cong 0.98$, and $\gamma \cong 0.67$ [36]. It is interesting that according to Shcherbakov et al. [80] this kind of distribution also holds for pure aftershock sequences, but with different parameters and exponents. In a recent analytical study of the ETAS model, Saichev and Sornette [74] obtained a different form for $f(x)$,

$$f(x) = \left[an\theta\rho^{\theta}x^{-1-\theta} + \left(1 - n + na\rho^{\theta}x^{-\theta}\right)^2\right] \exp\left[(n-1)x - \frac{na\rho}{1-\theta}x^{-1-\theta}\right], \tag{4.15}$$

and show that also this form, with $n = 0.9$, $\theta = 0.03$, $a = 0.76$ and $\rho = 1$ fits nicely Corral's data including the regime of very small x values. The parameter θ actually is not an independent fit parameter but describes how the Omori exponent p deviates from 1.

Very recently, we suggested an alternative approach to explain the observed form of $f(x)$, which is based on the assumption of long-term memory in the seismic activity [35]. We focused on the stationary parts of both Californian earthquake catalogues (Southern California Earthquake Catalogue (SCEC) and Northern California Earthquake Catalogue (NCSN) [81]) from 1995 to 1998 and find evidence that both magnitudes and return intervals are long-term correlated. We have shown explicitly that the long-term correlations can explain the fluctuations of magnitudes and return intervals and at the same time, without any fit parameter, the form of $f(x)$ in the whole x-regime, including the regime of very small values of x.

When searching for long-term memory in the relatively short stationary seismic data set, we compared the results for the observed data set with those predicted by a long term correlated simulated data set of the same length and the same distribution, where the value of the correlation exponent γ is obtained from the DFA analysis of the seismic record. To create an artificial data set we have taken Gutenberg–Richter distributed magnitudes above $M = -1$. The length of the data set was chosen such that the number of events above $M = 2$ and $M = 3$ are roughly the same as for the Californian catalogues. These magnitudes have been correlated by Fourier-filtering in combination with the Schreiber–Schmitz method. For a detailed description of these methods, we refer to [31,37–39]. The occurrence times of the events above $M = -1$ can either be chosen randomly or equidistant. Our results presented here do not depend on this choice, since the mean return interval between events $M \geq 2$ ($M \geq 3$) are 1 000 (10 000) times larger than the mean return interval between events $M \geq -1$.

Figure 4.20 compares the simulated record with $\gamma = 0.4$ with the seismic record of the SCEC between 1995 and 1998. Figures 4.20(a), (c) and (d) show the same part of the simulated data set for $M \geq -1$, $M \geq 2$ and $M \geq 3$. A subsequence of this part for $M \geq -1$ is shown in Fig. 4.20(b) where the characteristic mountain-valley structure of long-term correlated data which represents the clustering of large and small events in the record, can be clearly observed.

For $M \geq 2$ this structure becomes less pronounced, and is hardly seen for $M \geq 3$. Of course, the long-term correlations are also present in the subsets $M \geq 2$ or $M \geq 3$, but seem to play a less dominant role. Figures 4.20(e) and (f) show a part of the seismic record for $M \geq 2$ and $M \geq 3$. The sequence has been chosen such that the number of events is roughly the same in both the real and the simulated sequence. The seismic data look quite similar to the simulated ones. We will show in the next figures that the similarity is not only qualitatively (as in Fig. 4.20) but quantitatively, which strongly supports the hypothesis that seismic records are characterized by long-term correlations. For the quantitative comparison, we have considered (i) the records of consecutive magnitudes above $M = 2$ and above $M = 3$ and (ii) the records of consecutive return intervals between events above $M = 2$ and above $M = 3$, for both simulated ($\gamma = 0.4$) and observed records. In order to test for long-term correlations in these records, we have used DFA0 and DFA1 and calculated the corresponding fluctuation functions $F(s)$. For higher order of DFA we obtained similar results as presented here for DFA0 and DFA1 since we deal with stationary records.

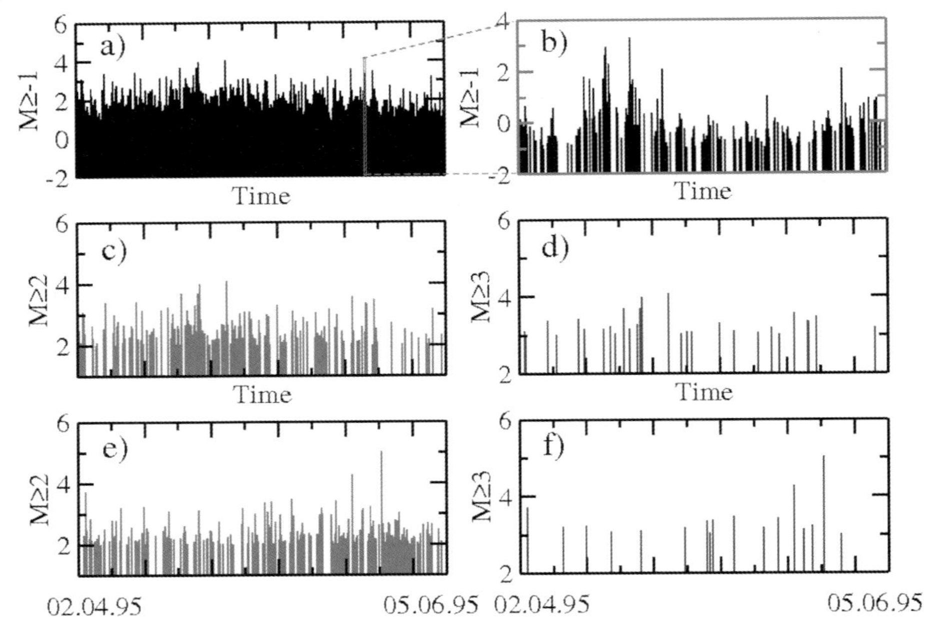

Fig. 4.20 (a–d) Sequence of artificial Gutenberg–Richter distributed correlated data with $\alpha = 0.8$. The magnitudes above $M = -1$ (a, b), $M = 2$ (c) and $M = 3$ (d) are shown. In (b), where only a small sector of the sequence (a) is displayed, one can clearly see the clustering of the events. (e–f) Sequence of real earthquake data (SCEC) with magnitudes above $M = 2$ (e) and $M = 3$ (f). The sequence has been chosen that the number of events above $M = 2$ and $M = 3$ are roughly the same as in the simulated record. One can also recognize here clustering of the events. (Taken from [35]).

Figure 4.21 shows the magnitude fluctuation functions $F(s)$ for (a) $M \geq 2$ and (b) $M \geq 3$. In both cases, the fluctuation functions in the double logarithmic plot are approximately straight lines. For both simulated and observed records, the slopes are around 0.59 for $M \geq 2$ and about 0.50 for $M \geq 3$. We have also calculated the fluctuation functions of the magnitudes $M \geq 2.5$. In this case the slope is around 0.56. Accordingly, simulated and observed records show the same quantitative correlation behavior. The fact that in the simulated data the exponents are smaller than the generated one ($\alpha = 0.8$) results from the fact that the consecutive events in the sub-records with $M \geq 2$ and $M \geq 3$ are far away in the original record (on the average separated by R_M) and thus appear to be less correlated since the extra noise screens the long-term correlations. To detect the real α-value one needs to look at much larger data sets where the asymptotic slope will approach α of the original sequence. This crossover effect increases with increasing M, and this explains why for $M \geq 3$ long-term correlations even seem to be absent in the simulated data set.

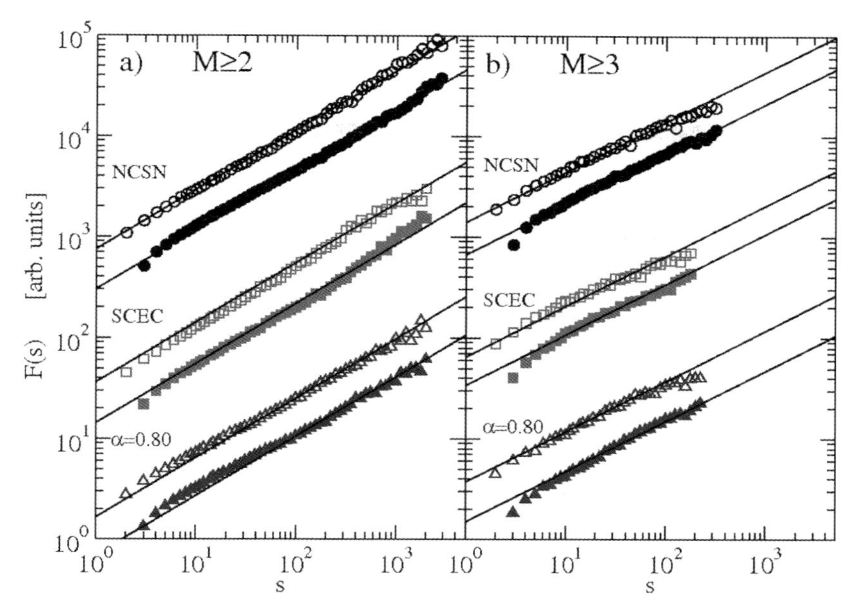

Fig. 4.21 Detrended fluctuation analysis of the magnitudes from the SCEC catalogue (1995–1998) (black circles), the NCSN catalogue (1995–1998) (squares) and an artificial Gutenberg–Richter distributed correlated data set with $\alpha = 0.8$ (triangles) for (a) $M \geq 2$ and (b) $M \geq 3$. The open symbols show the DFA0 results and the filled symbols show the DFA1 results. Both, DFA0 and DFA1, show an exponent around $\alpha = 0.59$ for $M \geq 2$ and around 0.50 for $M \geq 3$, the corresponding slopes are represented by straight lines in the figure. (Taken from [35]).

Figure 4.22 shows our test of the return intervals for long-term memory. Like in Fig. 4.21, the fluctuation functions in the double logarithmic plot are approximately straight lines that slightly bend up at large s-values. However, the slopes are larger than for the magnitudes, being close to 0.78 for $M = 2$ and 0.70 for $M = 3$, for both artificial and real records. We have also calculated the fluctuation functions of the return intervals for $M = 2.5$. In this case the slope is around 0.75, for both simulated and observed records. The fact that in the simulated data the exponents are smaller than the generated ones ($\gamma = 0.4$) has the same origin as explained above, but the effect of noise is smaller.

Accordingly, we have obtained quantitative agreement between the fluctuation functions of the observed seismic records and the simulated long-term correlated record with $\gamma = 0.4$. Figure 4.23 compares the distribution density $P_M(r)$ of the return intervals r for both kinds of records. After having identified the correlation exponent $\gamma = 0.4$ from Figs. 4.21 and 22 there is no free parameter to fit $P_M(r)$, since both length and distribution of the simulated data set are the same as for the observed data sets. Figure 4.23(a) shows an excellent agreement between the distributions of the observed and the simulated records plotted in scaled form. Figure 4.23(b) compares the distributions of the

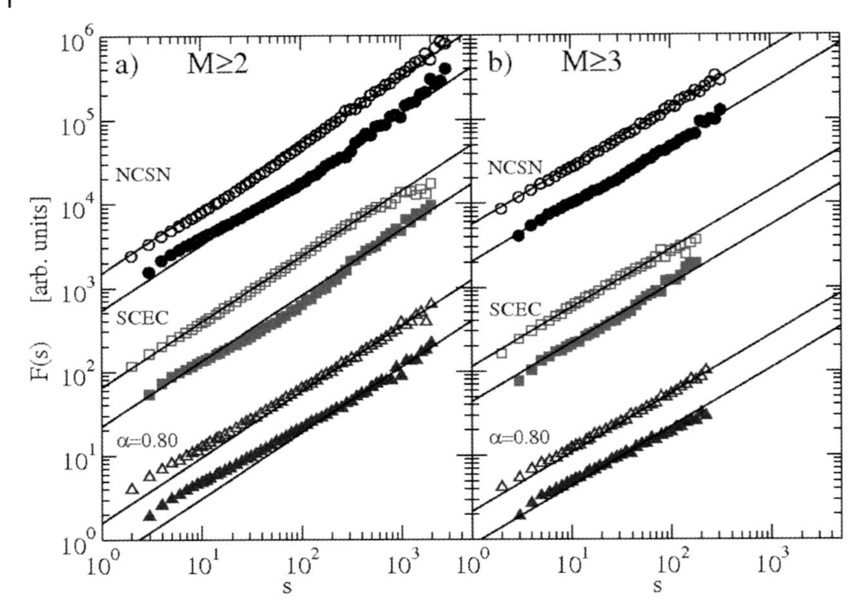

Fig. 4.22 Detrended fluctuation analysis of the return intervals from the SCEC catalogue (1995–1998) (black circles), the NCSN catalogue (1995–1998) (squares) and an artificial Gutenberg–Richter distributed correlated data set with $\alpha = 0.8$ (triangles) for (a) $M \geq 2$ and (b) $M \geq 3$. The open symbols show the DFA0 results and the filled symbols the DFA1 results. Both, DFA0 and DFA1, show an exponent around $\alpha = 0.78$ for $M \geq 2$ and around 0.70 for $M \geq 3$, the slopes are shown by straight lines in the figure. (Taken from [35]).

simulated record from Fig. 4.23(a) with the distributions for a large number of further earthquake data obtained by Corral [36]. The agreement is also excellent. As found earlier [31], the long-term correlated data scale over a large range of return intervals, but larger fluctuations occur at small ratios of r/R_M. It is remarkable that also in the observed data this feature can be seen.

4.5 Conclusions

In summary, we have studied the effect of long-term correlations in time series upon extreme value statistics and the return intervals between extreme events. For the return intervals in long-term persistent data with Gaussian, exponential, power-law, and log-normal distribution densities, we have shown that mainly the correlations rather than the distributions affect the return interval statistics, in particular the distribution density of return intervals, the conditional distribution density of return intervals, the conditional mean return intervals, and the correlation properties of the return interval series. The

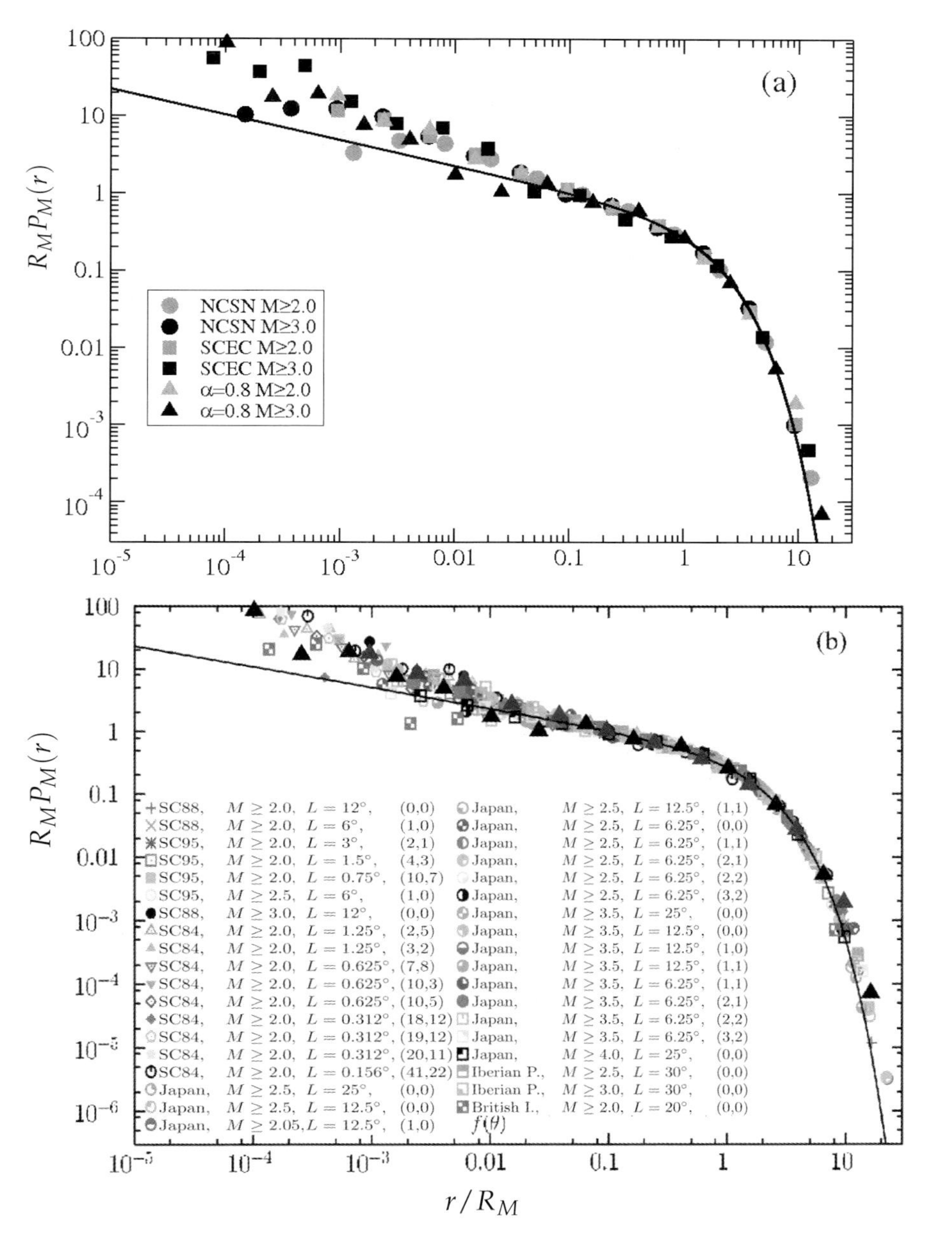

Fig. 4.23 Distribution density $P_M(r)$ of return intervals r from real earthquake data (a) SCEC and NCSN [81] and (b) obtained by Corral [36] in comparison with simulated correlated earthquake data with $\alpha = 0.8$ (triangles). $P_M(r)$ is shown for the magnitude thresholds listed in the inset. The black curve is Corral's fit given by Eq. (4.14). (Adapted from [35]).

stretched exponential decay of the return interval distribution density for long return intervals is complemented by a power-law decay for small return intervals, which will dominate the observable behavior in most (rather short) observational data. Still, the scaling behavior with the mean return interval holds in both regimes. We have also shown that the long-term persistence inherent in hydroclimate records represents a natural mechanism for the clustering of the hazardous events. We have found that, as a consequence of the long-term memory, the mean residual time to the next event increases with the elapsed time and depends strongly on the previous return interval. We have demonstrated that also this counterintuitive phenomenon can be seen in long climate records.

Furthermore, we studied seismic records in regimes of stationary seismic activity in Northern and Southern California. In summary, our results suggest that both magnitudes and return intervals (between seismic events above a certain threshold M) are long-term correlated with the same correlation exponent $\gamma = 0.4$. We consider it particular interesting that the long-term correlations can explain, without any fit parameter, the scaling form of the distribution function of the return intervals in the seismic records.

Considering series of maxima within segments of size R of the original data, we have shown numerically that the maxima distribution functions still converge to the same type of Gumbel distributions as for uncorrelated data for increasing R. For finite values of R, however, some deviations occur especially for originally Gaussian distributed data. Our extensive numerical simulations suggested that contrary to the common assumption in extreme value statistics, the maxima time series turn out to be *not* independently, identically distributed numbers. The series of maxima rather exhibit long-term correlations similar to those in the original data. Most notably we have found that the maxima distribution as well as the mean maxima significantly depend on the history, in particular on the value of the previous maximum.

Nevertheless, further work is needed to test if our findings are similar in other (non-Gaussian) initial distributions. In addition, we suggest that memory via the conditional mean maxima and conditional maxima distributions should be considered for improved risk estimation in long-term correlated data. It is also plausible that multiscaling, which occurs, e.g., in many hydrological time series, might have an even more significant impact on risk estimation and the prediction of extreme events like floods. Further work is definitely required to study the effects of multiscaling in time series upon extreme value statistics.

Acknowledgement

This work has been supported by the Bundesministerium für Bildung und Forschung of Germany and the Israel Science Foundation.

References

1 Bunde, A. and Havlin, S. (Eds.) (1994) *Fractals in Science*. Springer, Berlin.

2 Bunde, A., Kropp, J., and Schellnhuber, H.-J. (Eds.) (2002) *The Science of Disasters – Climate Disruptions, Heart Attacks, and Market Crashes*. Springer, Berlin.

3 Hurst, H.E., Black, R. P., and Simaika, Y. M. (1965) *Long-term Storage: An Experimental Study*. Constable & Co. Ltd., London.

4 Mandelbrot, B. B., and Wallis, J. R.(1969) Wat. Resour. Res. 5, 321–340.

5 Rodriguez-Iturbe, I., and Rinaldo, A. (1997) *Fractal River Basins – Change and Self-Organization*. Cambridge University Press.

6 Kantelhardt, J. W., Koscielny-Bunde, E., Rybski, D., Braun, P., Bunde, A., and Havlin, S. (2006) J. Geophys. Res. (Atmosph.) 111, 1106.

7 Koscielny-Bunde, E., Kantelhardt, J. W., Braun, P., Bunde, A., and Havlin, S. (2006) J. Hydrol. 322, 120–137.

8 Koscielny-Bunde, E., Bunde, A., Havlin, S., and Goldreich, Y. (1996) Physica A 231, 393–396.

9 Pelletier, J. D., and Turcotte, D. L. (1997) J. Hydrol. 203, 198–208.

10 Koscielny-Bunde, E., Bunde, A., Havlin, S., Roman, H. E., Goldreich, Y., and Schellnhuber, H.-J. (1998) Phys. Rev. Lett. 81, 729–732.

11 Talkner, P., and Weber, R. O. (2000) Phys. Rev. E 62, 150–160.

12 Eichner, J. F., Koscielny-Bunde, E., Bunde, A., Havlin, S., and Schellnhuber, H.-J. (2003) Phys. Rev. E 68, 046133.

13 Shlesinger, M. F., West, B. J., and Klafter, J. (1987) Phys. Rev. Lett. 58, 1100–1103.

14 Prasad, R. R., Meneveau, C., and Sreenivasan, K. R. (1988) Phys. Rev. Lett. 61, 74–77.

15 Peng, C.-K., Buldyrev, S. V., Havlin, S., Simons, M., Stanley, H. E., and Goldberger, A. L. (1994) Phys. Rev. E 49, 1685–1689.

16 Peng, C.-K., Buldyrev, S. V., Goldberger, A. L., Havlin, S., Sciortino, F., Simons, M., and Stanley, H. E. (1992) Nature 356, 168–170.

17 Arneodo, A., Bacry, E., Graves, P. V., and Muzy, J. F. (1995) Phys. Rev. Lett. 74, 3293–3296.

18 Bunde, A., Havlin, S., Kantelhardt, J. W., Penzel, T., Peter, J.-H., and Voigt, K. (2000) Phys. Rev. Lett. 85, 3736–3739.

19 Kantelhardt, J. W., Penzel, T., Rostig, S., Becker, H. F., Havlin, S., and Bunde, A. (2003) Physica A 319, 447–457.

20 Liu, Y. H., Cizeau, P., Meyer, M., Peng, C.-K., and Stanley, H. E. (1997) Physica A 245, 437–440.

21 Yamasaki, K., Muchnik, L., Havlin, S., Bunde, A., and Stanley, H. E. (2005) PNAS 102 (26), 9424–9428.

22 Leland, W. E., Taqqu, M. S., Willinger, W., and Wilson, D. V. (1994) IEEE/ACM Transactions on networking 2, 1–15.

23 Paxson, V. and Floyd, S. (1995) IEEE/ACM Transactions on networking 3, 226–244.

24 Meko, D. M., Therrell, M. D., Baisan, C. H., and Hughes, M. K. (2001) J. Amer. Wat. Resour. Assoc. 37 (4), 1029–1039.

25 Whitcher, B., Byers, S. D., Guttorp, P., and Percival, D. B. (2002) Wat. Resour. Res. 38 (3), 1054.

26 Mann, M. E., Bradley, R. S., and Hughes, M. K. (1999) Geophys. Res. Lett. 26, 759–762.

27 Moore, J. J., Hugen, K. A., Miller, G. H., and Overpeck, J. T. (2001) J. Paleolimnology 25, 503–517.

28 Grissino-Mayer, H. D. (1996) *A 2129-Near Annual Reconstruction of Precipitation for Northwestern New Mexico*, USA. In: J. S. Dean, D. M. Meko, and T. W. Swetnam (Eds.), *Tree Rings, Environment, and*

Humanity. Radiocarbon 1996, Dept. Geosciences, University of Arizona, Tucson, pp. 191–204.

29 Bunde, A., Eichner, J. F., Kantelhardt, J. W., and Havlin, S. (2005) Phys. Rev. Lett. 94, 048701.

30 Kantelhardt, J. W., Koscielny-Bunde, E., Rego, H. H. A., Havlin, S., and Bunde, A. (2001) Physica A 295, 441–454.

31 Eichner, J. F., Kantelhardt, J. W., Bunde, A., and Havlin, S. (2007) Phys. Rev. E 75, 011128.

32 Bunde, A., Eichner, J. F., Kantelhardt, J. W., and Havlin, S. (2003) Physica A 330, 1–7.

33 Bunde, A., Eichner, J. F., Kantelhardt, J. W., and Havlin, S. (2004) Physica A 342, 308–314.

34 Eichner, J. F., Kantelhardt, J. W., Bunde, A., and Havlin, S. (2006) Phys. Rev. E 73, 016130.

35 Lennartz, S., Livina, V. L., Bunde, A. and Havlin S. (2007) *Long-term Memory in Earthquakes and the Distribution of Interoccurrence Times*. Submitted to Phys. Rev. Lett.

36 Corral, A. (2004) Phys. Rev. Lett. 92, 108501.

37 Turcotte, D. L. (1992) *Fractals and Chaos in Geology and Geophysics*. Cambridge University Press.

38 Makse, H. A., Havlin, S., Schwartz, M., and Stanley, H. E. (1996) Phys. Rev. E 53, 5445–5449.

39 Schreiber, T., and Schmitz, A. (1996) Phys. Rev. Lett. 77, 635–638.

40 Altmann, E. G., and Kantz, H. (2005) Phys. Rev. E 71, 056106.

41 Kac, M. (1947) Bull. Amer. Math. Soc. 53, 1002–1010.

42 Storch, H. v., and Zwiers, F. W. (2001) *Statistical Analysis in Climate Research*. Cambridge Univ. Press.

43 Oliveira, J. G., and Barabasi, A.-L. (2005) Nature 437, 1251–1253.

44 Vazquez, A., Oliveira, J. G., Dezsö, Z., Goh, K.-I., Kondor, I., and Barabasi, A.-L. (2006) Phys. Rev. E 73, 036127.

45 Sornette, D., and Knopoff, L. (1997) Bull. Seism. Soc. Am. 87, 789–798.

46 Gumbel, E. J. (1958) *Statistics of Extremes*. Columbia University Press, New York.

47 Galambos, J. (1978) *The Asymptotic Theory of Extreme Order Statistics*. John Wiley & Sons, New York.

48 Leadbetter, M. R., Lindgren, G., and Rootzen, H. (1983) *Extremes and Related Properties of Random Sequences and Processes*. Springer, New York.

49 Galambos, J., Lechner, J., and Simin, E. (Eds.) (1994) *Extreme Value Theory and Applications*. Kluwer, Dordrecht.

50 Embrechts, P., Klüppelberg, C., and Mikosch, T. (1997) *Modelling Extremal Events*. Springer, Berlin.

51 Fisher, R. A., and Tippett, L. H. C. (1928) Proc. Camb. Phil. Soc. 24, 180–190.

52 Chow, V. te (1964) *Handbook of Applied Hydrology*. McGraw-Hill Book Company, New York.

53 Raudkivi, A. J. (1979) *Hydrology*. Pergamon Press, Oxford.

54 Rasmussen, P. F. and Gautam, N. (2003) J. Hydrol. 280, 265–271.

55 Antal, T., Droz, M., Györgyi, G., and Racz, Z. (2001) Phys. Rev. Lett. 87, 240601.

56 Bramwell, S. T., Holdsworth, P. C. W., and Pinton, J.-F. (1998) Nature 396, 552–554.

57 Bramwell, S. T., Christensen, K., Fortin, J.-Y., Holdsworth, P. C. W., Jensen, H. J., Lise, S., Lopez, J. M., Nicodemi, M., Pinton, J.-F., and Sellitto, M. (2000) Phys. Rev. Lett. 84, 3744–4747.

58 Dahlstedt, K. and Jensen, H. J. (2001) J. Phys. A: Math. Gen. 34, 11193–11200.

59 Dean, D. S. and Majumdar, S. N. (2001) Phys. Rev. E 64, 046121.

60 Raychaudhuri, S., Cranston, M., Przybyla, C., and Shapir, Y. (2001) Phys. Rev. Lett. 87, 136101.

61 Majumdar, S. N. and Comtet, A. (2004) Phys. Rev. Lett. 92, 225501.

62 Majumdar, S. N. and Comtet, A. (2005) J. Stat. Phys. 119, 777–826.

63 Guclu, H. and Korniss, G. (2004) Phys. Rev. E 69, 65104(R).

64 Berman, S. M. (1964) Ann. Math. Statist. 35, 502–516.

65 Moberg, A., Sonechkin, D. M., Holmgren, K., Datsenko, N. M., and Karlén, W. (2005) Nature 433, 613–617. Data from: http://www.ndcd.noaa.gov/paleo/recons.html

66 Turcotte, D. L., (1995) Proc. Natl. Acad. Sci. USA 92, 6697.

67 Turcotte, D. L., and Schubert, G., (2001) *Geodynamics*, Cambridge University Press.

68 Kagan, Y. Y., (1999) Pure Appl. Geophys. 155, 233.

69 Omori, F., and Coll. J., (1894) Sci. Imp. Univ. Tokyo 7, 111.

70 Gutenberg, B., and Richter, C. F., (1941), Geol. Soc. Am. Spec. Pap. 34, 1–131.

71 Bath, M., (1965), Tectonophysics 2, 483–514.

72 Helmstetter, A. and Sornette, D., (2002) J. Geophys. Res. 107, 2237.

73 Saichev, A. and Sornette, D., (2006) Eur. Phys. J. B. 49, 377–401.

74 Saichev, A. and Sornette, D., (2006) Phys. Rev. Lett. 97, 078501.

75 Bak, P., Christensen, K., Danon, L., and Scanlon, T., (2002) Phys. Rev. Lett. 88, 178501.

76 Corral, A., (2003) Phys. Rev. E 68, 035102(R).

77 Corral, A., (2005) Phys. Rev. E 71, 017101.

78 Davidsen, J., and Goltz, C., (2004) Geophys. Res. Lett. 31, L21612.

79 Tosi, P., de Rubeis, V., Loreto, V., and Pietronero, L., (2004) Annals of Geophys. 47 (4), 1849.

80 Shcherbakov, R., Yakovlev,G., Turcotte, D. L., and Rundle, J. B., (2005) Phys. Rev. Lett. 95, 218501.

81 Northern California Seismic Network (NCSN, 1995–1998) – http://quake.geo.berkeley.edu/ncedc/catalog-search.html; Southern California Earthquake Center (SCEC, 1995–1998) – www.data.scec.org/catalog-search.html

5
Factorizable Language: From Dynamics to Biology

Bailin Hao and Huimin Xie

There is no universal measure of complexity. When the problem under study leads to description by means of symbolic sequences, formal language theory may provide a convenient framework for analysis. In this review we concentrate on a special class of languages, namely, factorizable languages to be defined later, which occur in many problems of dynamics and biology. In dynamics we have in mind symbolic dynamics of unimodal maps and complexity of cellular automata. In biology we draw examples from DNA and protein sequence analysis.

5.1
Coarse-Graining and Symbolic Description

Let us start by making the following observation [1].

A high-energy physicist recognizes the six lower case letters u, d, c, s, b, t as quark names and associates them with a certain mass, charge and quantum numbers such as "charm" or "flavor". More scientists use the symbols p, n, e to denote proton, neutron and electron each having a certain mass, charge, spin or magnetic moment, but they are not concerned with from which three quarks a proton or neutron is made.

Chemists consider $H, C, N, O, P, S, \cdots$ to be element names and know their atomic number, ion radius, chemical valence and affinity. Chemical compounds may be denoted by combined use of such symbols as H_2O, NO, CO_2, and so forth. However, when it comes to writing chemical formulas for the nucleotides and amino acids which are the constituents of DNAs and proteins, there is no need to write down the tens of atomic symbols each time.

Biochemists call the nucleotides a, c, g, t and denote the amino acids by $A, C, \cdots, W, Y$. Now all one has to know is c and g are strongly conjugated by three hydrogen bonds while the weak coupling of a and t is made by two hydrogen bonds. Here "strong" and "weak" differ by many orders of magnitudes from that in high-energy physics. In a biochemical pathway or a meta-

Review of Nonlinear Dynamics and Complexity. Edited by Heinz Georg Schuster

ISBN: 978-3-527-40729-3

bolic network, proteins/enzymes are denoted by simple names and there is no need to spell out the amino acids that make the proteins.

This observation can be continued further. What is the moral learned? In describing Nature one cannot grasp the details on all levels at the same time; one has to concentrate on a particular level at a time treating larger scales as background and reflecting smaller scales in "parameters". For example, in describing the Brownian motion of pollen the environment at large is represented by a temperature while the friction force is given by using a coefficient of friction. If necessary, one could go down to the molecular level to calculate the coefficient directly. This is called coarse-grained description. Coarse-graining is reached by making "approximations", that is, by ignoring details on finer scales. Nevertheless, it may lead to rigorous conclusions. Geoffrey West, the President of the Santa Fe Institute, once made a remark that had Galileo be equipped with our high-precision measuring instruments he would not be able to discover the law of free falling bodies and would have to write a 42-volume *Treatise on Falling Bodies*.

Furthermore, coarse-grained description of Nature is always associated with the use of symbols. If one is lucky enough these symbols may form symbolic sequences. Coarse-grained description of dynamics leads to symbolic dynamics [2]. Biochemists represent DNA and proteins as symbolic sequences. It is an essential fact that all these sequences are one-dimensional, directed and unbranching chains made of letters from a finite alphabet, thus bringing these sequences into the realm of language theory.

Since we have come to the notion of symbolic sequences, it is appropriate to recollect a basic fact on huge collections of symbolic sequences. In Shannon's seminal 1948 paper [3] that laid the foundation of modern information theory, besides the famous definition of information now familiar to all students, he stated a few other Theorems. Theorem 3 in [3] can be roughly interpreted as follows. Given a sequence of length N made of 0s and 1s, there are in total 2^N such sequences. Generally speaking, when N gets very large, these 2^N sequences can be divided into two subsets: a huge subset of "typical" sequences and a small group of "atypical" sequences. The statistical property of a typical sequence resembles that of any other typical sequence or the bulk of the huge group, while the property of any atypical sequence is very specific and has to be scrutinized almost individually. The simplest members of the atypical set are sequences made of N consecutive 1s or 0s as well as various kinds of periodic and quasi-periodic sequences. However, the most significant ones from the atypical set are those with hidden regularities mixed with a seemingly random background. These are the truly complex sequences we have to characterize. While the typical set may be characterized by statistical means, the atypical sequences require more specific methods to explore them, including combinatorics, graph theory and formal language theory.

One should not be misled by the adjective "formal". Given the right context, language theory may provide a framework for a rigorous description of complexity and a workable scheme for down-to-number calculation of characteristics.

5.2 A Brief Introduction to Language Theory

Basic notions of formal language theory may be found in many monographs, for example, [4,5], and in the comprehensive handbook on formal languages [6]. Therefore, we only give a brief account of some basic notions.

5.2.1 Formal Language

We start from an alphabet Σ made of a finite number of letters. For example, $\Sigma = \{0,1\}$ or $\{L,R\}$ in symbolic dynamics of unimodal maps [2]; $\Sigma = \{A,C,G,T\}$ when dealing with DNA sequences; in studying protein sequences Σ consists of the 20 single-letter symbols for the amino acids. Collecting all possible finite strings made of letters from the alphabet Σ and including an empty string ϵ, we form a set Σ^*. The empty string ϵ contains no symbol at all, but plays an important role in language theory. Its presence makes formal languages a kind of monoid in algebra [7]. The collection of all non-empty strings is denoted by Σ^+. Obviously, $\Sigma^* = \Sigma^+ \bigcup \{\epsilon\}$.

Now comes the definition of a formal language: any subset $L \subset \Sigma^*$ is called a language. With such a general definition one cannot go very far. The key point is how to specify the subset L. A powerful way to define formal languages makes use of generative grammar. A subset of the alphabet Σ is designated as initial letters. Then a collection of production rules is applied repeatedly to the initial letters and to strings thus obtained. All strings generated in this way form the language L.

Obviously, all strings in the complementary set $L' = \Sigma^* - L$ are inadmissible in the language L. We reserve the term *forbidden word* for members of a special subset of L':

Definition 5.1 A word $x \in \Sigma^+$ is called a *forbidden word* of language L if $x \notin L$ but every proper substring of x belongs to L.

The set of all forbidden words of a language L is denoted by L''. The set L'' may be defined for any language L and may be empty. However, it will play a key role in the study of a special class of languages, namely, factorizable languages, to be defined later in Section 5.2.2.

N. Chomsky classified all possible sequential production rules in the mid 1950s and defined four classes of languages as briefly summarized in Table 5.1. In the order of increasing complexity these classes are regular language (RGL), context-free language (CFL), context-sensitive language (CSL) and recursively enumerable language (REL). Each class of language corresponds to a class of automata with different memory requirement.

Table 5.1 The Chomsky hierarchy of languages and automata.

Language	Corresponding automaton	Memory
Regular	Finite state automaton (FA)	Limited
Context-free	Push-down automaton	A stack
Context-sensitive	Linearly bounded automaton	Proportional to input
Recursively enumerable	Turing machine	Unlimited

As seen from Table 5.1, each language class corresponds to a class of automata. The finite state automata (FA) that recognize RGL are especially instructive because they may be constructed explicitly in many cases. If a FA has a designated starting state and every state has unambiguous transitions to other states upon reading in a symbol it is a deterministic FA (DFA); otherwise it is called a non-deterministic FA (NDFA). DFA and NDFA are equivalent in their capability to recognize the appropriate language. An NDFA may be transformed into a DFA by means of "subset construction". Among all DFAs accepting the same language there exists one with a minimal number of states. It is called a minimal DFA (minDFA). The size of the minDFA is determined by the index of an equivalence relation R_L generated by the language L in Σ^*. We will give an example of constructing a minDFA in Section 5.5.4. All related notions may be found in standard textbook like [4].

In 1968 the developmental biologist A. Lindenmayer introduced parallel production rules to study the growth of simple multicellular organisms. This approach developed into another framework for the classification of formal languages – the Lindenmayer system or L-system. There are more classes in the L-system: D0L (Deterministic no interaction), 0L and IL (non-deterministic no interaction and with interaction); if the production rules are chosen from a set of rules called a Table the languages become T0L and TIL. In the Chomsky system symbols are divided into terminal and non-terminal symbols, the latter being working variables that do not appear in the final products. The L-system has been later extended to include non-terminal symbols to make E0L, ET0L and EIL languages. Referring the interested readers to the monograph [8], we show the relation of various languages in Fig. 5.1.[1]

1) We take this opportunity to correct an inexactitude in the original figure on p. 389 of [2].

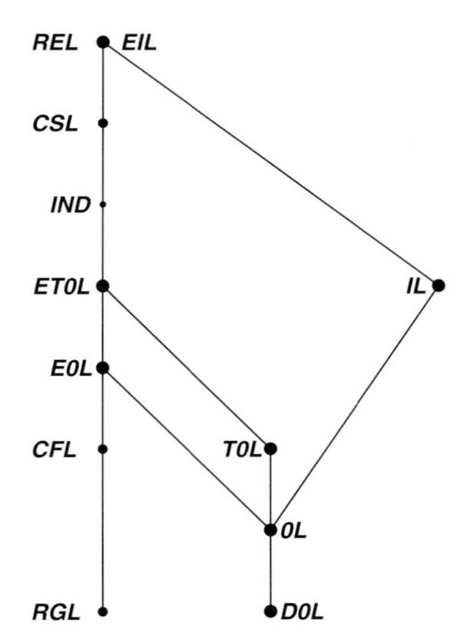

Fig. 5.1 The relation between Chomsky hierarchy and the L-system. IND denotes *indexed language* not discussed here.

5.2.2 Factorizable Language

In this review we concentrate on factorizable languages which appear in dynamics and some biological applications but have not been mentioned in standard textbooks like [4] or the handbook [6]. First, the definition:

Definition 5.2 A language L has *factorial property* or L is called a *factorizable language* if from $x \in L$ it follows that all substrings of x including the empty string ϵ belong to L.

A factorizable language L is determined by the set L'' of forbidden words:

$$\begin{aligned} L &= \Sigma^* - L' &= \Sigma^* - \Sigma^* L'' \Sigma^*, \\ L' &= \Sigma^* - L &= \Sigma^* L'' \Sigma^*. \end{aligned} \tag{5.1}$$

In fact, it follows from the factorizability of L that any string $x \notin L$ must contain at least one forbidden word; contrariwise, putting any number of letters in front or behind a forbidden word must lead to strings not contained in L. A member of the set L'' was called a *Distinct Excluded Block* (DEB) in Wolfram's analysis of grammatical complexity of cellular automata [9].

In order to calculate L'' from L' we need two operators MIN and R acting on any language M and defined in formal language theory [4]:

$$\begin{aligned} \mathrm{MIN}(M) &= \{x \in M \,|\, \text{no proper prefix of } x \text{ is in } M\}, \\ \mathrm{R}(M) &= \{x \,|\, x^R \in M\}, \end{aligned}$$

where x^R means the mirror of x obtained by reversing the string x. Then we have

$$L'' = R \circ MIN \circ R \circ MIN(L'). \tag{5.2}$$

We sketch the proof of Eq. (5.2) by considering a word $x \in L'$. Inspecting x from the start until encountering a forbidden word $v \in L''$, we may write $x = uvz$ where v is the only forbidden word in uv and z denotes all the rest of x after the first forbidden word v. In fact, z may be any string $w \in \Sigma^*$ and uvw represents any member of L'. Now MIN(L') leaves only strings of form uv where $v \in L''$. Further operations are simple: R $\circ$ MIN(L') produces strings of form $v^R u^R$, MIN $\circ$ R $\circ$ MIN(L') strips the latter to v^R and an additional R restores v. Since v may be any member of L', Eq. (5.2) holds.

However, Eq. (5.2) is rarely needed in practice when the set of forbidden words L'' is known beforehand. Then the language is determined directly from $L = \Sigma^* - \Sigma^* L'' \Sigma^*$ as we shall see in subsequent sections.

5.3 Symbolic Dynamics

Symbolic dynamics [2] arises in coarse-grained descriptions of dynamics. Generally speaking, a dynamics f maps the phase space X into itself:

$$f : X \to X.$$

Suppose X is a compact space then there exists a finite covering of X. We label each covering by a letter from a finite alphabet Σ. By ignoring the precise location of a point $x \in X$ and recording only the label of the corresponding covering, the action of the dynamics f corresponds to a shift $\mathcal{S}$ in the space of all strings over the alphabet Σ:

$$\mathcal{S} : \Sigma^* \to \Sigma^*.$$

The shift dynamics $\mathcal{S}$ acting on the space of symbolic sequences over the alphabet Σ makes a symbolic dynamics. In this general setting the notion of symbolic dynamics applies to one-dimensional as well as multi-dimensional dynamics, and to conservative as well as dissipative systems. Further specification of f and X enriches the symbolic dynamics as we have learned from symbolic dynamics of one- and two-dimensional mappings and ordinary differential equations [2].

5.3.1 Dynamical Language

Symbolic sequences occurring in a symbolic dynamics may be viewed as a language L. Any symbolic sequence (word) generated by the dynamics is admissible in the dynamical language L. It was recognized in the early study of symbolic dynamics [10] that a dynamical language L has the following two properties:

1. Factorizability: any substring of a word in L also belongs to L, because any part of a longer symbolic sequence is also generated by the same dynamics. This property alone makes L a factorizable language.
2. Prolongability: any word in L may be appended by a letter from Σ to get another word in L due to the time development of the dynamics. Even a "non-moving" fixed point corresponds to adding one and the same letter repeatedly to prolong the symbolic sequence.

5.3.2 Grammatical Complexity of Unimodal Maps

In one-dimensional discrete dynamical systems the unimodal maps of the interval and circle maps are the best-studied examples. Since these cases have been discussed at length in monographs such as [2] and [5], we only summarize the new knowledge on grammatical complexity of languages in symbolic dynamics of unimodal maps as of the end of 1990s in Fig. 5.2 [11] and Table 5.2.

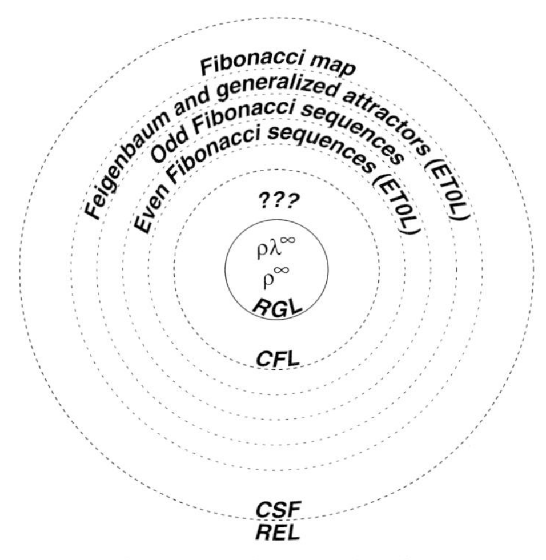

Fig. 5.2 Grammatical complexity of languages in symbolic dynamics of unimodal maps as of the end of the 1990s.

Table 5.2 Forbidden words in regular languages of unimodal maps.

Kneading sequence	**Set of forbidden words** L''	**Remark**
RL^{∞}	Empty set $\emptyset$	A surjective map
L^{∞}	$\{R\}$	Fixed point only
R^{∞}	$\{RL\}$	Period 2 appears
$(RL)^{\infty}$	$\{RLL, RLRR\}$	Period 4 appears
$(RLR)^{\infty}$	$\{RLL\}$	Period 3 appears
RLR^{∞}	$\{RL(RR)^n L\}_{n\geq 0}$	First band-merging point

A few explanations follow. Symbolic dynamics of unimodal maps uses two letters R and L to denote the **R**ight and **L**eft parts of the unit interval divided by the critical point C of the map. A special symbolic sequence corresponding to the numerical trajectory starting from the first iterate of C is called a *kneading sequence*. Kneading sequences serve as "topological" or universal parameters of the map that do not depend on the concrete shape of the unimodal mapping function. Among the new results obtained in the 1990s we indicate:

1. In the dynamical languages of the unimodal maps the class of regular languages contains only periodic kneading sequences ρ^{∞} and eventually periodic kneading sequences $\rho\lambda^{\infty}$ [12], where ρ and λ are finite strings made of R and L. Examples of these sequences are given in Table 5.2.

2. Since the above result [12] closes the problem of regular languages in unimodal maps we draw a solid circle around RGL in Fig. 5.2.

3. All attempts to construct context-free languages associated with unimodal maps, including various Fibonacci sequences [13], have led to context-sensitive languages which are not context-free. Hence the as yet open conjecture: there is no context-free language which is not regular in unimodal maps [5], a fact represented by the three question marks in Fig. 5.2.

4. The kneading sequence of the infinite limit of the Feigenbaum period-doubling cascade may be obtained by infinitely repeated application of the homomorphism $h = (R \rightarrow RL, L \rightarrow RR)$ to the single letter R. It was the only known context-sensitive language in the unimodal maps in the beginning of the 1990s, first proved to be a non-context-free language [14], in which the language consists of the set of all substrings of the kneading sequence only. Now the class of context-sensitive language has split into layers as shown in Fig. 5.2. All these layers are not empty and may be characterized precisely in mathematical terms [15].

5.4 Sequences Generated from Cellular Automata

The research of cellular automata originated from Von Neumann's study of formalizing the self-reproduction feature of life in the 1950s [16], and popularized by The Game of Life invented by Conway in the 1970s [17, Chap. 25]. The systematic study of cellular automata, however, was begun in 1980s in a series of works by S. Wolfram et al., in which many new points of view and tools were introduced. It is evident that the ideas and results from nonlinear science play an important role in cellular automata, including numerical experiment by computer, statistical methods, algebraic methods and the method of formal languages and automata from the theoretical computer science.

Two collections of papers are good references for an introduction into the area of cellular automata [18, 19]. For the arguments caused by Wolfram's new book [20] in 2002 many materials can be found from various websites, for example, http://www.math.usf.edu/~eclark/ANKOS_reviews.html.

In this section we will consider the study of sequences generated by one-dimensional cellular automata. In Section 5.4.1 an introduction is given, including some definitions and simple facts about cellular automata. Sections 5.4.2 and 5.4.3 are respectively devoted to two kinds of complexity of sequences generated by cellular automata, namely limit complexity and evolution complexity. The languages involved in these studies are always factorizable languages.

5.4.1 Encounter the Cellular Automata

The complex systems in Nature are often composed by coupling many simple systems and *Cellular Automata (CA)* are one of ideal mathematical models for those complex systems.

The definition of one-dimensional CA can be given as follows[2].

Assume that there are infinite cells, namely automata, situated on all integer points of the number axis, each cell can only take finite states, and the time variable is discrete, namely, $t = 0, 1, 2, \ldots$. It seems true that these discrete features of states and time are similar to those features of many automata in theoretical computer science [4]. The distinct feature of CA, however, is that here there are infinitely many cells, and they change their states simultaneously at each discrete time. In addition, the state of each cell at the next time is determined by the states of itself and some cells nearby through some definite rules specified by a given CA, and these rules are all the same for each cells and each time thenceforth.

2) Both CA in [16, 17] are two-dimensional.

Hence the evolutionary rules of CA are homogeneous both for space and time. From the point of view of computing, CA belong to a class of parallel computers.

Now the above definition can be rephrased by mathematical language as follows. Assume that each cell has $k > 1$ states, and denote these states by the symbols $0, 1, \ldots, k-1$, hence the alphabet set is $S = \{0, 1, \ldots, k-1\}$. Let $a_i \in S$ be the state of the cell posited at the integer $i \in \mathbb{Z}$, the set of all integers, and the state of CA at each time be a bisequence over S:

$$a = \cdots a_{-n} \cdots a_{-2} a_{-1} a_0 a_1 a_2 \cdots a_n \cdots , \tag{5.3}$$

which is called a *configuration* of CA, and $S^{\mathbb{Z}}$, the set of all bisequence, the configuration space of CA.

Furthermore, if the states of cells at time t are denoted by a_i^t, $i \in \mathbb{Z}$, then the evolutionary rule of states can be written in the form of

$$a_i^{t+1} = f(a_{i-r}^t, \cdots, a_i^t, \cdots, a_{i+r}^t), \quad \forall i \in \mathbb{Z}, \tag{5.4}$$

in which the number r is called the radius of neighborhood. The existence of such r reflects the finiteness of information transmission velocity in CA.

The simplest CA are those CA with the number of states $k = 2$ and the radius of neighborhood $r = 1$, which are conventionally called *elementary cellular automata (ECA)*.

Since for ECA the alphabet set being $S = \{0, 1\}$, the rule f in the definition of (5.4) is a mapping from S^3 to S. As the arguments of this mapping have only eight possibilities, namely, $000, 001, \cdots, 111$, hence an ECA is completely given as long as the values of f are given on these eight arguments. A consequence of this consideration is that there are altogether $2^{2^3} = 256$ possible ECA.

A popular coding scheme for ECA can be explained below (see [18]).
For example, let an ECA be given by the rules

$$011, 100, 101 \rightarrow 1; \quad 000, 001, 010, 110, 111 \rightarrow 0, \tag{5.5}$$

namely $f(011) = 1$, $f(100) = 1$, and so forth. Rearrange the rules (5.5) as follows

111	110	101	100	011	010	001	000
0	0	1	1	1	0	0	0

and then convert the binary number 00111000 in the second line to the decimal number 56, and call the ECA thus given the ECA of rule 56.

It is convenient to see the space-time behavior of one-dimensional CA on a computer screen. Figure 5.3 shows the experimental results performed for the space-time behavior of eight different ECA.

In Figure 5.3 the rule numbers of each ECA are put on the top of each subfigure. Every subfigure is obtained by the same procedure as follows. Let the small white square be the symbol 0, and the small black square the symbol 1. The first line is (a part of) the starting configuration at the time $t = 0$, which is generated by a pseudorandom binary generator. The direction of time is from top to bottom. Each of the other lines is obtained from the previous line by the rule of ECA. For each subfigure of Fig. 5.3 the number of time steps is 100, and the length of the part of the starting configuration is 300. Since only the center part of width 100 is seen, the problem of the appropriate boundary effect is removed. (Another experiment method not discussed here for one-dimensional CA is to use the circular boundary condition, and it can be seen as if all cells are arranged on a circle.)

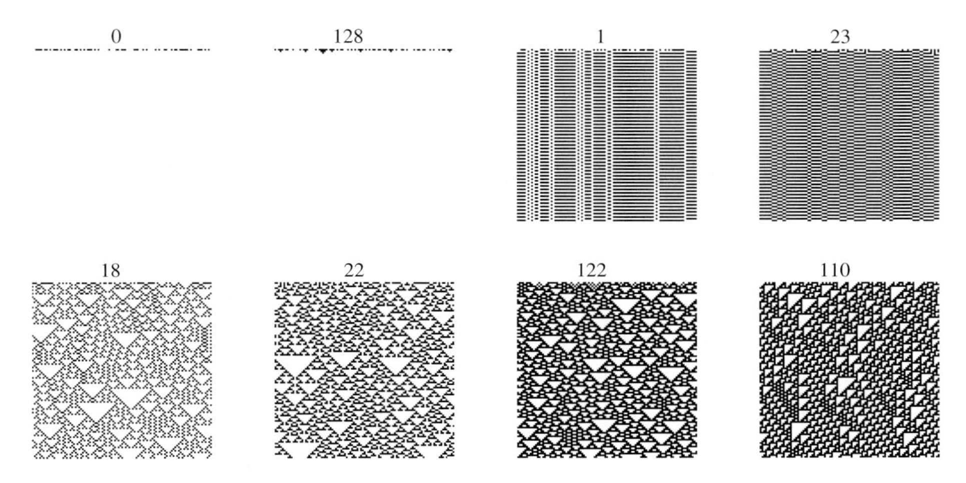

Fig. 5.3 The space-time behavior of eight ECA of different rules.

Wolfram performed many computer experiments for one-dimensional and two-dimensional CA, including all 256 ECA, and proposed a classification scheme for all CA as follows:

1. all configurations converge to an invariant homogeneous configuration composed of all the same symbols,
2. converge to some simple stable or periodic structures,
3. chaotic and non-periodic behaviors appeared,
4. complex local structures appeared, and some of them can irregularly propagate.

As pointed by Wolfram all these four kinds of behaviors can be seen in ECA.

Of course this classification scheme is only a phenomenological one, and more rigorous analysis is absolutely needed. In Fig. 5.3 all four kind of behaviors are shown, but there exist many questions which cannot be removed by pure experiment on a computer.

For example, the first ECA of rule 0 in Fig. 5.3, namely the mapping from S^3 to the state 0, belongs to the class 1, since each configuration from $t = 1$ composed by the same symbol 0s, which will be denoted by $\overline{0}$ hereafter. But it is more difficult for the next one, the ECA of rule 128, since nearly any experiment with this ECA shows that the configuration will rapidly converge to $\overline{0}$ as for the ECA of rule 0, but can we say definitely that the ECA of rule 128 belongs to class 1 in Wolfram's classification scheme? We will come back to this question in Section 5.4.2.

Another factor is that different starting configurations can lead to different space-time behaviors. An example is shown in Fig. 5.4. The left subfigure is obtained from a random starting configuration, and the behavior looks like those in the class 2. The starting configuration for the middle subfigure is a random sequence of 00 and 11, and the behavior looks like those in the class 3. The right subfigure is a combination of those of the two previous subfigures.

Fig. 5.4 Space-time behavior of ECA of rule 94 for different starting configurations.

There are many studies for ECA, but still it cannot said that any complete theory has been established. An example is the ECA of rule 110 in Fig. 5.3. It has been believed that it belongs to the class 4, and only recently an important result has been obtained in that the ECA of rule 110 has the power of a universal Turing machine [20,21].

The remaining part of this Section is devoted to a review of research of sequences generated from CA by the method of formal languages.

5.4.2 Limit Complexity of Cellular Automata

The introduction of the method of formal languages to CA began mainly from Wolfram's paper [9], in which many conjectures were made based on computer experiments. After that some theoretical results were obtained by Hurd [22,23], in which and also in other papers the concept of limit set plays an important role. The complexity analyzed by this method will be called the *Limit Complexity of CA*.

First define both the limit set of CA and the limit language generated from the limit set.

The rule f in Eq. (5.4) is also called the *local mapping* of CA. Using f onto a configuration will generate a new configuration, and hence establish a *global mapping* F from the configuration space into itself. We also use the notation, for example, f_{56} and F_{56} to represent the local and global mapping for the ECA of rule 56.

Using F iteratively generates a dynamical system, and the configuration space, $S^{\mathbb{Z}}$, becomes the phase space for this dynamical system. The difference between this dynamical system generated by the global mapping of CA and the low-dimensional discrete dynamical systems is that the former one is an infinite-dimensional dynamical systems. Nevertheless, many mature concepts from dynamical systems can be borrowed for the study of CA, for example, invariant sets, periodic points, trajectories, and so forth. An important feature of CA is that we have to consider its space structure with its time structure simultaneously, namely, the space-time behaviors.

Consider the phase space $S^{\mathbb{Z}}$ more closely. Since $S^{\mathbb{Z}}$ is a product space generated from the alphabet set S, considering the discrete topology to S, and using the Tychonoff theorem, $S^{\mathbb{Z}}$ is a compact space. Moreover, this compact topology can also be metrizable, for example, by introducing the distance between $x, y \in S^{\mathbb{Z}}$ as follows: let $d(x, y) = 0$ if $x = y$, and

$$d(x,y) = \frac{1}{k+1},$$

where the number k is the minimal non-negative integer such that $x_k \neq y_k$ or $x_{-k} \neq y_{-k}$. Since it can be shown that the global mapping F is continuous on the space $S^{\mathbb{Z}}$, the image set of $F(S^{\mathbb{Z}})$ is also a compact set.

From the inclusion relation $F(S^{\mathbb{Z}}) \subset S^{\mathbb{Z}}$ it is evident that $F^{n+1}(S^{\mathbb{Z}}) \subset F^n(S^{\mathbb{Z}})$ holds for every $n \geq 0$. By the theorem about the nonemptyness of the intersection of non-increasing sequence of compact sets, the set

$$\Lambda(F) = \bigcap_{n=0}^{\infty} F^n(S^{\mathbb{Z}}) \tag{5.6}$$

is non-empty, and called the *limit set* of CA hereafter. It can be seen that the limit set contains all periodic points and non-wandering points of F.

From the point of view of dynamical systems, it is evident that the limit set is just the largest invariant set of the global mapping F. But the limit set of CA still has its special features as shown by the following theorem [24, p.373]:

Theorem 5.3 *If the limit set of a CA has more than one element, then it must be an infinite set.*

The simplest case is the limit set which has only one element: $\Lambda(F) = \{c\}$. Since any shift of element in limit set is still an element of it, the configuration c must be a space homogeneous one, namely composed by a single symbol q. Considering that the configuration c must also be an invariant point of F, then it is certain that a rule of the form $qq \cdots q \rightarrow q$ must be in f of Eq. (5.4). We call F a *nilpotent CA* if its limit set contains only one element. It corresponds to the class 1 in Wolfram's classification scheme. But it turns out that even here some difficulty is unavoidable. It is proved already that the problem of deciding whether a given CA is nilpotent is an undecidable problem, that is to say there exists no algorithm to give the answer of "yes" or "no" for the problem of deciding whether a given CA is nilpotent or not. This result has been proved for some time for CA with more than one dimension, and for one-dimensional CA lately in [25].

In order to reduce the study of bisequences to the study of finite sequences, a language will be generated from the limit set.

For a subset A of the compact space $S^{\mathbb{Z}}$, taking all finite subsequences for each bisequence of A gives a formal language over S denoted by the notation $\mathcal{L}(A)$. It can be shown that, if the set A is shift-invariant and closed, then it can be determined by the language $\mathcal{L}(A)$ completely [5].

The rest of this section is devoted to the study of the limit language of CA, and its complexity in the Chomsky hierarchy [4].

The first result in this aspect is the following theorem. We will give its proof, and refer the reader to the textbook [4] for the concepts and tools used below.

Theorem 5.4 *If F is the global mapping of a CA, then for each $n \geq 0$, the language $\mathcal{L}(F^n(S^{\mathbb{Z}}))$ is regular.*

Proof. Since for each n the mapping F^n is also a CA, it suffices to prove the conclusion for the case of $n = 1$. The method used below is to construct the finite automaton accepting the language $\mathcal{L}(F(S^{\mathbb{Z}}))$ exactly.

Let the radius of neighborhood of F be r, and the number of states be k. Taking all words of length $2r$ as the accepting states of finite automaton, then there are k^{2r} states altogether.

The transition rule of states can be determined as follows. Observing the local mapping f as shown in Eq. (5.4), every rule

$$f(s_1 s_2 \cdots s_{2r+1}) = s_0$$

can be interpreted as a transition from the state $s_1 s_2 \cdots s_{2r}$ to the state $s_2 s_3 \cdots s_{2r+1}$ if the symbol s_0 is read. In other words, between these two states there exists an arc labeling by s_0 as shown by

$$\boxed{s_1 s_2 \cdots s_{2r}} \xrightarrow{s_0} \boxed{s_2 s_3 \cdots s_{2r+1}}.$$

It is easy to see that this finite automaton accepts exactly the language $\mathcal{L}(F(S^{\mathbb{Z}}))$. □

The limit language has a similar expression to Eq. (5.6):

$$\mathcal{L}(\Lambda(F)) = \bigcap_{n=0}^{\infty} \mathcal{L}(F^n(S^{\mathbb{Z}})). \tag{5.7}$$

Hence from Theorem 5.4 it is known that the limit language is the intersection of countable regular languages, but it cannot tell us how complex the limit language is. In Hurd's papers [22,23] many CA are constructed such that their limit languages can have all possible grammatical complexity in Chomsky hierarchy.

From the expression of Eq. (5.7) it is a consequence that a sequence $s \in \mathcal{L}(\Lambda(F))$ if and only if $s \in \mathcal{L}(F^n(S^{\mathbb{Z}}))\,\forall\, n = 1, 2, \cdots$, and verifying this is usually a hard task, if not impossible.

Now consider the limit languages of ECA. As discussed above that all nilpotent CA belong to the class 1 in Wolfram's scheme, and as shown in Fig. 5.3 the ECA of rule 0 is one of them. But what about the ECA of rule 128 in Fig. 5.3? Here the computer experiment can easily lead to the conjecture that this ECA is also one of class 1, since each experiment shows that it converges to all 0's rapidly. Examing its local mapping

111	110	101	100	011	010	001	000
1	0	0	0	0	0	0	0

means that the configuration of all 1s also belongs to the limit set, and from Theorem 5.3 the latter must be an infinite set.

It is not hard to tackle directly the limit language of ECA of rule 128, and obtain the result that

$$\mathcal{L}(\Lambda(F_{128})) = 0^*1^*0^* = \{0^m 1^n 0^p \mid m, n, p \geq 0\}, \tag{5.8}$$

a regular language. Since it is a factorizable language, the concept of forbidden words (or distinct excluded blocks) [5,26] can be used and it is easy to find out all forbidden words for this language (5.8) being $\{10^n 1 \mid n \geq 1\}$.

In [9] a measure of complexity for regular languages is proposed and applied to characterize the behavior of CA. For a regular language L, define the *regular language complexity* of L as the number of states of *the minimal deterministic finite automaton* (minDFA) which accepts L. In [18] the regular language complexity of $\mathcal{L}(F^n(S^{\mathbb{Z}}))$ for $n = 1 \sim 5$ and 256 ECA are computed and listed as Table 10 of it.

It turns out that for most ECA belonging to the class 1 or 2, their regular language complexity is either constant or increasing slowly; but for ECA belonging to the class 3 or 4, their regular language complexity is increasing rapidly such that in many cases the computation of regular language complexity cannot be performed even for $n = 4$ or 5.

But this method is still not a rigorous and reliable approach to decide the complexity level of limit languages. In Table 5.3 these results are listed for some ECA whose complexity is already known.

Table 5.3 Regular language complexity of some ECA for $n = 1 \sim 5$

ECA	$n = 1$	$n = 2$	$n = 3$	$n = 4$	$n = 5$
94	15	230	3904		
22	15	280	4506		
122	15	179	5088		
104	15	265	2340	1394	1542
164	15	116	667	1214	
110	5	20	160	1035	

The limit language of the ECA of rule 94 is proved to be non-regular [27] and this result coincides with the data in Table 5.3. However, we can compare these conclusions with the space-time behavior of ECA of rule 94 shown in Fig. 5.4, in which the left subfigure is the behavior obtained from a random starting configuration, and it seems that the configuration converges to a period 6 attractor. This comparison reveals that sometimes the experimental method is not satisfactory and reliable.

The conclusion about the ECA of rule 22 is similar and proved in [28], but its space-time behavior as shown in Fig. 5.3 is more complex than that of ECA of rule 94.

It can be seen from Fig. 5.3 that the space-time behavior of the ECA of rule 122 is similar to that of ECA of rule 22. Here the theoretical analysis gives a better result in that its limit language, $\mathcal{L}(\Lambda(F_{122}))$, is not a context-free language [29].

If it is true that the data in Table 5.3 give the correct evidence of complexity for the first three ECA, those of rules 94, 22 and 122, then it is not so for the next two ECA, those of rules 104 and 164. As proved in Jiang's doctoral thesis, both limit languages of these two ECA are regular, although their regular language complexity are very high indeed. However, seeing their data in Table 5.3, it seems that both ECA of rules 104 and 164 may be more complex than the ECA of rule 110, but it is known that the latter is the most complex ECA as proved in [20, 21].

The most unsuccessful cases of the application of limit languages happen for the surjective CA, in which it is evident that

$$F(S^{\mathbb{Z}}) = S^{\mathbb{Z}}, \quad \Lambda(F) = S^{\mathbb{Z}}, \quad \mathcal{L}(\Lambda(F)) = S^*. \tag{5.9}$$

That is to say the configuration space $S^{\mathbb{Z}}$ itself, namely, the phase space, is invariant under the global mapping F, and the limit language is the largest language over the alphabet S. Its regular language complexity is simply 1, since there is only one accepting state as it can accept every sequence over S.

There are 30 ECA belonging to this category, and some of them are shown in Fig. 5.5.

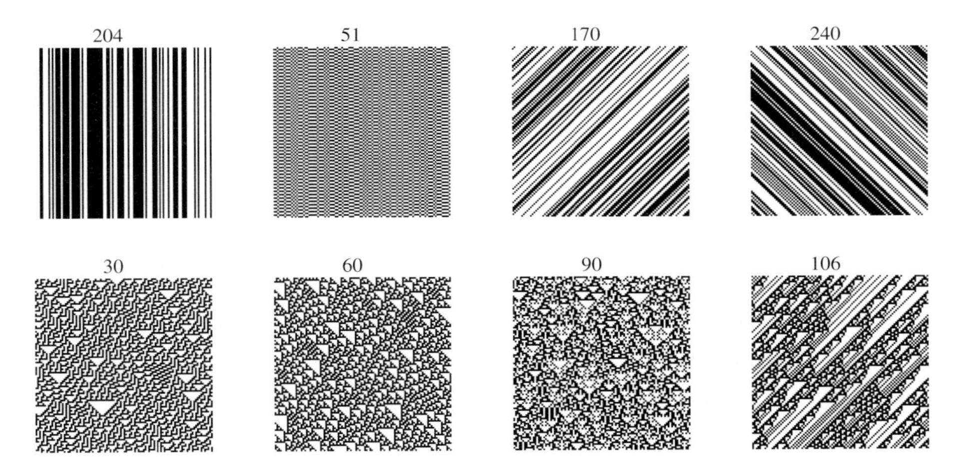

Fig. 5.5 The space-time behaviors of eight surjective ECA.

It is evident that the four surjective ECA in the first line of Fig. 5.5 are very simple. The ECA of rule 204 is the identity mapping, each configuration is a fixed point of it; and the ECA of rule 51 is the flip-flop mapping, its f_{51} maps 0 to 1 and vice versa, each configuration is a period-2 point of it. The real radius of neighborhood of these two ECA is $r = 0$. It is true that both of them belong to the class 2 in Wolfram's classification scheme. The ECA of rules 170 and 240 are respectively the left shift and right shift.

It is also evident that the four surjective ECA in the second line of Fig. 5.5 show much more complex behaviors than those ECA of the first line and belong to the class 3 in Wolfram's classification scheme. The ECA of rule 30, among them, is used as a pseudorandom number generator in the software of Mathematica, and there are many discussions about the behaviors of ECA of rules 60 and 90. The ECA of rule 106 contains the leftshift ECA of rule 170 as its factor system, but is much more complex than it.

Summary

Although there exist many results and discussions about the limit set and limit language of CA, no universal method or result has been found. Computer experiments are useful for one-dimensional CA, but the results need careful exploration and theoretical research.

The classification problem of CA has drawn much attention after Wolfram's scheme, and several rigorous classification schemes have been obtained [30, 31]. However, there are often drawbacks in these schemes that (1) most complex CA are classified into one class, and not much help is provided for their analysis, (2) there is no effective criteria to determine which class a given CA belongs to.

It seems that there are still many open problems in the area of CA. It is also true even of ECA, the simplest class of CA.

5.4.3 Evolution Complexity of Cellular Automata

From the content of Section 5.4.2 it seems that the approach using the limit language of CA has its restriction, and is especially unsatisfactory for the surjective CA.

Many examples, which include the above-mentioned surjective CA, show that there exists no direct relationship between the limit set (or limit language) and the space-time behavior of CA. It is not surprising if we recall the fact that the limit set is just the largest invariant set of the dynamical system of CA.

Another approach for analyzing the complexity of CA using formal languages is proposed by Gilman [32]. The complexity analyzed by this approach will be called the *Evolution Complexity of CA* hereafter.

The evolution complexity considers the evolution of one cell, a_i, or some cells, $a_i, a_{i+1}, \ldots, a_{i+k-1}$, in which case we say its width is k. From the space homogeneous feature of CA, the index i can be fixed as 0 or some other number. Therefore, this approach is very similar to looking at the computer screen for the CA's evolution with a fixed width of cells, that is a fixed width of window, and it can be expected that the evolution complexity of the CA should have a more direct relationship with the space-time behavior of CA obtained by computer experiment.

Considering that the configuration of CA is a bisequence, which is infinite in both directions, it is natural that people can only see its finite part and take the evolution of this finite strip as the basis for analyzing its complexity. This idea is another example of coarse-graining mentioned before.

Assuming the width of the observing window is k, then the object of our study can be seen as a sequence over the alphabet set S^k. Taking all possible finite sequences together gives the *evolution language of width k* of the CA, which will be denoted by the notation $\mathcal{E}_k$. For the case of $k = 1$ the name *time series* of CA is also used to describe the sequence over S, which is a record of one cell's evolution over some time interval.

For a given CA and a given sequence s over S^k, it can be determined whether s belongs to $\mathcal{E}_k$ by using the local rule (5.4) for a fixed number of cells of CA, and this number is a linear function of k. Therefore, this is a problem which can be solved by a restricted kind of Turing machine, namely, the linear bounded automata. Using the relation between automata and Chomsky hierarchy a theorem is obtained that the grammatical complexity of the evolution language will not be beyond the *context-sensitive languages* [4,32,33].

An example of CA is given by Gilman, in which the number of states is $k = 2$, and the radius of neighborhood is $r = 2$, hence it is not an ECA. The local rule of this CA is given by

$$a_i^{t+1} = f(a_{i-2}^t a_{i-1}^t a_i^t a_{i+1}^t a_{i+2}^t) = a_{i+1}^t a_{i+2}^t \quad \forall i \in \mathbb{Z}, \tag{5.10}$$

where the right-hand side is the multiplication of states a_{i+1}^t and a_{i+2}^t which are binary numbers as well as symbols.[3)] The space-time behavior of Gilman's CA is shown in Fig. 5.6.

Fig. 5.6 The space-time behavior of Gilman's CA.

It has been proved that the evolution language $\mathcal{E}_1$ of Gilman's CA is a context-sensitive language, but not a context-free language (see [32, p. 99] and [33, p. 429]).

The time series also appears in [34,35], in which the problem of periodicity of time series of ECA is discussed, in which $f(000) = 0$ is satisfied and there exist only finite symbol 1 in the starting configurations.

For the ECA of rule 18, the complexity of its evolution languages of each width has been completely solved. In [36] it is proved that for this ECA the language $\mathcal{E}_1$ is regular, and all $\mathcal{E}_k$ ($k \geq 2$) are context-sensitive, but not context-free. Similar results were obtained recently for the ECA of rule 146 [37].

For the ECA of rule 22, it is proved that all $\mathcal{E}_k$ ($k \geq 2$) are not regular [38], but it is open for $\mathcal{E}_1$.

3) Although the CA of Gilman has the radius of neighborhood $r = 2$, but its a_i^{t+1} is determined only by a_{i+1}^t and a_{i+2}^t, and, therefore, this CA can be obtained by the composition of the ECA of rule 136 and rule of 170, the left-shift operator.

Comparing the limit complexity and the evolution complexity of the same CA, there exists no direct relationship between them. For example, the evolution complexity of Gilman's CA is the highest level possible in Chomsky hierarchy, but its limit language is very simple, since it can be proved that for Gilman's CA, $\mathcal{L}(\Lambda(F)) = 0^*1^*0^*$, the same language as that of ECA of rule 128 mentioned before.

Another example in this aspect is the ECA of rule 90. As pointed out above, it is a surjective CA, which has been studied extensively before. We know that its limit language is $\{0,1\}^*$, its regular language complexity is 1, and cannot reflect its rather complex space-time behavior as shown in Fig. 5.5. Generally, the ECA of rule 90 is considered a typical CA in class 3 of Wolfram's classification scheme.

It is easy to show that for the ECA of rule 90, each $\mathcal{E}_k$ $(k \geq 1)$ is regular. Let their regular language complexity be $C(\mathcal{E}_k)$, then it can be verified that

$$C(\mathcal{E}_1) = C(\mathcal{E}_2) = 1, \ C(\mathcal{E}_k) = 2^{k-2} + 2 \, \forall \, k \geq 3, \tag{5.11}$$

in which the first two results are equivalent to saying that

$$\mathcal{E}_1 = \{0,1\}^*, \quad \mathcal{E}_2 = \{00,01,10,11\}^*.$$

As an example of the results in Eq. (5.11), the minDFA accepting the evolution language $C(\mathcal{E}_3)$ of the ECA of rule 90 will be constructed below.

Figure 5.7 shows a DFA which contains three accepting states q_0, q_1 and q_2, and one non-accepting state q_3.

The alphabet set of the language $C(\mathcal{E}_3)$ is the set of all words of length 3, namely,

$$S = \{000,001,010,011,100,101,110,111\},$$

and its elements are denoted by $0 \sim 7$ in Fig. 5.7. The local mapping of ECA of rule 90 is

$$f(a_{-1}, a_0, a_1) \equiv a_{-1} + a_1 \pmod 2,$$

and it is a straightforward verification that the automaton in Fig. 5.7 is the required one accepting $C(\mathcal{E}_3)$.

In the sequel a recent work is reviewed in which the evolution complexity of width 1 for all 256 EAC is explored and some interesting results obtained [39,40].

Since the evolution language $\mathcal{E}_1$ is factorizable, it can be characterized by its forbidden words [5,26]. Using a C++ program as a searching tool, we can use a computer to find all the forbidden words whose lengths are not beyond

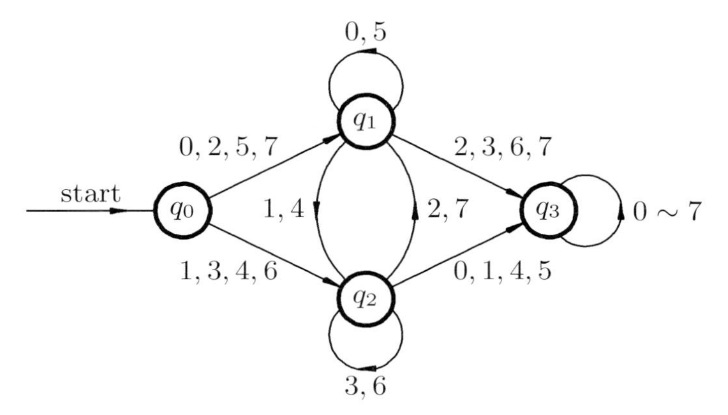

Fig. 5.7 A minDFA accepting the evolution language of width 3 of the ECA of rule 90.

a certain limitation, then a theoretical analysis is followed to see whether the whole set of forbidden words can be determined.

Since the complexity of space-time behavior is not influenced by the exchange of 0, 1 and the mirror image of the neighborhood, the 256 ECA can be reduced to 88 classes of ECA, and each class thus formed is represented by the ECA in the class with the minimal rule number [18]. Hence, we need only to consider 88 ECA.

The result is shown in Table 5.4, in which all 88 ECA are classified into four classes according the complexity of $\mathcal{E}_1$ for each ECA.

Class I includes all ECA whose $\mathcal{E}_1$ has no forbidden words at all, namely, their $\mathcal{E}_1 = S^*$, and the above-mentioned ECA of rule 90 belongs to this class. Moreover, Class I is divided into two Subclasses I.1 and I.2, the difference is that every ECA in I.1 is surjective, but the ECA in I.2 is not.

Table 5.4 The classification of ECA on the basis of complexity of evolution languages $\mathcal{E}_1$

Class	No.	Rule number of ECA	No. of FW	Complexity
I.1	10	15, 30, 45, 60, 90, 105, 106, 150, 154, 170	none	$\{0,1\}^*$
I.2	4	94, 122, 126, 184	none	$\{0,1\}^*$
II	53	0, 1, 2, 3, 4, 5, 6, 8, 10, 11, 12, 13, 14, 18, 19 23, 24, 29, 32, 34, 35, 36, 38, 40, 42, 43, 44 46, 50, 51, 57, 58, 72, 76, 77, 108, 128, 130 132, 136, 138, 140, 142, 146, 152, 160, 162 168, 172, 178, 200, 204, 232	≤ 5	FCR
III	8	27, 28, 33, 78, 104, 134, 156, 164	infinite	ICR
IV	13	7, 9, 22, 25, 26, 37, 41, 54, 56, 62, 73, 74, 110	infinite	nonRGL

FW, forbidden words; FCR, finite complement regular language; ICR, infinite complement regular language; nonRGL, non-regular language.

The ECA in Class II has only finite forbidden words, and the number of them is in the range of $1 \sim 5$. Hence their evolution language $\mathcal{E}_1$ is finite complement regular, and it can be noted that the symbolic flow associated with them is of the *subshift of finite type (SFT)* [5, p. 23].

In Class III the evolution language of each ECA has infinite forbidden words, but still is regular, and, therefore, their $\mathcal{E}_1$ is infinite complement regular. The associated symbolic flow is the so-called *sofic system* [5, p. 25].

Finally, the theoretical study of the ECA in Class IV has not been finished yet; for some of them the evolution language $\mathcal{E}_1$ is proved to be context-free, but not regular, and for others we have only some feeling that their evolution languages $\mathcal{E}_1$ are non-regular, and maybe much more complex than those ECA which are known already.

This conjecture is supported by the data in Table 5.5, in which the numbers of forbidden words whose length are not beyond 17 are listed.

Table 5.5 The numbers of forbidden words of length $K = 2 \sim 17$ for ECA in Class IV

Rule	2	3	4	5	6	7	8	9	10	11	12	13	14	15	16	17	Sum
7	0	0	3	2	0	2	0	2	0	2	0	2	0	2	0	2	17
9	0	1	1	2	0	2	3	3	3	3	3	4	7	6	14	10	62
22	0	0	1	1	0	0	1	1	2	5	1	7	26	54	78	153	330
25	0	1	1	1	0	3	2	3	2	2	4	3	5	2	14	15	58
26	0	1	0	0	1	0	0	0	2	2	4	7	16	19	35	44	131
37	0	0	0	3	4	5	3	10	9	23	20	22	32	36	37	66	270
41	0	1	0	0	0	1	1	1	6	2	14	30	51	90	146	253	596
54	0	1	1	0	0	1	4	2	0	4	6	15	18	34	32	62	180
56	0	1	1	0	0	1	0	0	2	0	0	5	0	0	14	0	24
62	0	1	0	1	2	4	4	3	6	5	6	6	10	3	16	12	79
73	0	0	2	2	1	2	2	1	2	3	3	8	11	13	18	27	95
74	0	0	3	0	1	2	1	0	0	0	1	2	4	7	11	16	48
110	0	0	0	1	3	3	5	6	12	16	17	38	42	73	112	198	526

Finally, a theorem about the evolution language of the ECA of rule 56 in Class IV is cited below, which shows that an interesting structure appears in the time series generated by this ECA [40].

Theorem 5.5 *Let $\mathcal{E}_1$ be the evolution language of width 1 of the ECA of rule 56, then (1) the languages $\mathcal{E}_1$ and $\mathcal{E}_1''$, the set of its forbidden words, are both context-free, but not regular; (2) the set $\mathcal{E}_1''$ can be expressed explicitly by*

$$\mathcal{E}_1'' = \{111\} \cup 0D01,$$

in which D is the Dyck language generated by the strings 0 and 01; (3) the number of forbidden words whose length is $3n + 1$ ($n \geq 2$) is C_{n-1}, the $n - 1$-th Catalan number.[4)]

5.5
Avoidance Signature of Bacterial Complete Genomes

DNAs are one-dimensional, directed, non-branching heteropolymers made of four kinds of monomers – the nucleotides adenine (a), cytosine (c), guanine (g), and thymine (t). In 1995 the first two complete genomes of free-living bacteria were published. By the end of 2006 more than 430 bacterial genomes were available in public databases. Having a genome at one's disposal means that many global questions may be asked. A biochemist would wish to infer all possible metabolic pathways underlying the life of a particular bacterium. A physicist without sound biological knowledge might ask the simplest global question concerning the distribution of short nucleotide strings of a fixed length K, in particular, whether some strings are absent at a given K.

5.5.1
Visualization of Long DNA Sequences

The length of a typical bacteria genome is a few millions nucleotides. In order to visualize the K-string composition of a genome we apply a simple counting algorithm. To count the number of K-strings we allocate the 4^K counters on the computer screen as a direct product of K copies of the 2×2 matrix M:

$$M \otimes M \otimes \cdots \otimes M,$$

where

$$M = \begin{pmatrix} g & c \\ a & t \end{pmatrix}.$$

We use 16 colors to represent the counts. If a string is absent the corresponding cell is shown in white. The bright colors are assigned to small counts. If the counts are greater than a certain threshold, say, 40, the cell is shown in black.

4) The Dyck language is one of the most important context-free languages [4, p. 142], and the Catalan numbers appear in many interesting problems of combinatorics, their expression is given by

$$C_n = \frac{1}{n+1}\binom{2n}{n}, \; n \geq 1.$$

The first Catalan numbers are $1, 2, 5, 14, 42, 132, 429, \cdots$ [41].

This is also a kind of coarse-graining. A program entitled SEEDNA has been put in the public domain [42]. We call the output of this program a "portrait" of the bacterium.

5.5.2
Avoided *K*-Strings in Complete Genomes

A portrait of the harmless laboratory strain K12 of *Escherichia coli* is shown in Fig. 5.8 for $K = 8$. The almost regular pattern seen in Fig. 5.8 tells the under-representation of strings containing *ctag* as substring. In fact, closely related species have similar pattern of under-represented strings and bacteria from different taxa show some characteristic "avoidance signature". For more discussion see [43]. A portrait is nothing but a two-dimensional histogram of the string counts. The string counts may be visualized by using a one-dimensional histogram as well. The latter may show some peculiar fine structure for a few randomized genomes. The explanation requires a combination of simple combinatorics with statistics [44].

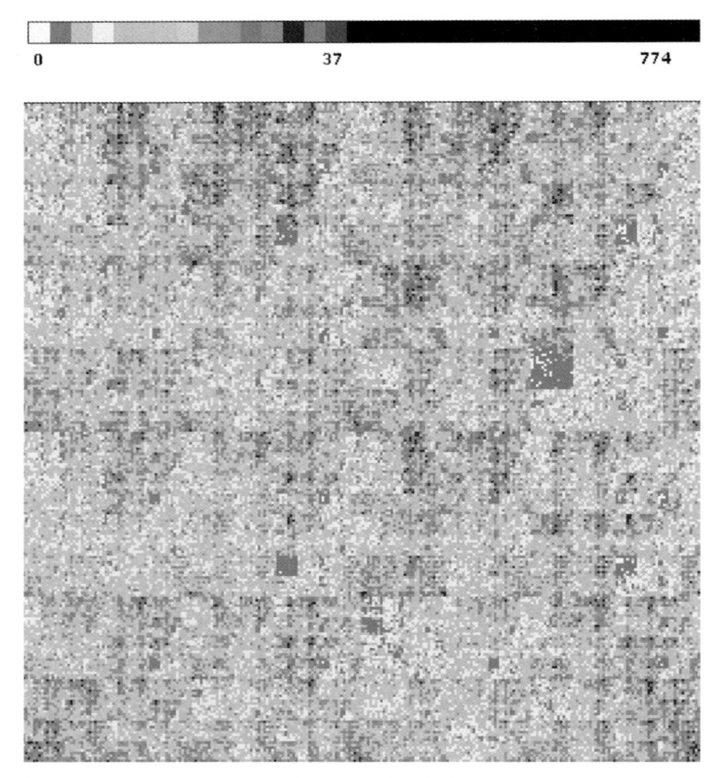

Fig. 5.8 A "portrait" of *E. coli* strain K12 at $K = 8$.

5.5.3
True and Redundant Avoided Strings

By inspection of Fig. 5.8 or, to be more precise, by direct counting, we see that at $K = 8$ there are 173 string types missing in the *E. coli* K12 genome. At $K = 7$ there is only one missing string, namely, *gcctagg*. This simple fact raises a question. Among the 173 missing strings at $K = 8$ eight strings must be the consequence of the string *gcctagg* being absent at $K = 7$, because one may add a letter in front or at the end of the string to get eight strings that cannot appear at $K = 8$. We say that at $K = 8$ there are eight redundant and 165 true missing strings in the genome. Given that at length K there is one missing string, one would like to know how many redundant missing strings it produces at $K + i$. By mathematical induction one gets a simple formula as the answer:

$$N_i = 4^i(i + 1), \quad i = 0, 1, \cdots \tag{5.12}$$

However, this formula only gives an approximate answer to the question as it has not taken into account the fact that the first and the last letters in the string *gcctagg* happen to be the same. In this particular case the above formula works for all $i < 13$, but at $i \geq 13$ it fails because the 13-string *gcctaggcctagg* contains the missing 7-string twice, a fact not reflected in the inductive derivation of Eq. (5.12).

The situation becomes more formidable when at certain K there are several true missing strings, and among the missing ones there exist overlaps among their prefixes and suffixes. A prominent example is provided by the hyperthermophilic bacterium *Aquifex aeolicus* genome [45]. In this 1 551 335 letter sequence four true missing strings are identified at $K = 7$:

$$B = \{gcgcgcg, gcgcgca, cgcgcgc, tgcgcgc\}. \tag{5.13}$$

The overlapping among these forbidden words is apparent. Denote by a_K the number of words of length K within Σ^* that do not contain any element of the subset B and define a generating function

$$f(s) = \sum_{K=0}^{\infty} a_K s^K, \tag{5.14}$$

where s is an auxiliary variable. An explicit expression of $f(s)$ may be obtained [46, 47] by invoking the Goulden–Jackson cluster method [48] in combinatorics. The Goulden–Jackson method is capable of determining the number of strings including or excluding designated substrings with overlapping prefixes and suffixes among the later. It works even for letters with non-equal probabilities of appearance [49]. Not going into the details of the combinatorics, we turn to the language theory solution of the same problem.

5.5.4 Factorizable Language Defined by a Genome

Given a complete genome G one may define a language as follows. Take enough copies of the same genome G and cut them in all possible ways, from single nucleotides, dinucleotides, trinucleotides, up to the uncut sequences themselves in G. The collection of all these strings plus an empty string ϵ defines a language $L(G) \subset \Sigma^*$ over the alphabet $\Sigma = \{a, c, g, t\}$. Clearly, the language $L(G)$ is factorizable by construction.

Now we show how formal language theory may provide a framework to yield a concrete solution to an appropriate problem in numbers. First of all, any language $L \subset \Sigma^*$ introduces an *Equivalence Relation* R_L in Σ^* with respect to L: any two elements $x, y \in \Sigma^*$ are equivalent and denoted as xR_Ly if and only if for every $z \in \Sigma^*$ both xz and yz either belong to L or not belong to L. As usual, the index of R_L is the number of equivalent classes in Σ^* with respect to L. An equivalent class may be represented by any element $x \in L$ of that class and we will denote this equivalent class by $[x]$. The importance of the equivalent relation R_L comes from the Myhill–Nerode Theorem in language theory, see, for example, reference [4]: L is regular if and only if the index of R_L is finite and the number of states in the minDFA that accepts L is given by the index.

Taking the four true missing strings in the *A. aeolicus* genome as forbidden words, that is, let $L'' = B$, we undertake to construct a finite state automaton that accepts $L(G)$. We define a set V:

$$V = \{v | v \text{ is a proper prefix of some } y \in L''\}.$$

Then for each word $x \in L$ there exists a string $v \in V$ such that it is equivalent to x, or using our notations xR_Lv. In other words, all equivalent classes of Σ^* with respect to L are represented in V. Therefore, in order to find all equivalent classes of Σ^* with respect to L it is enough to work with L''. By the way, $[\epsilon]$ and L' are always two equivalent classes among others.

Collecting all proper suffixes of the avoided strings in B, we get

$$\begin{aligned} V = & \{g, gc, gcg, gcgc, gcgcg, gcgcgc, c, cg, cgc, cgcg, \\ & cgcgc, cgcgcg, t, tg, tgc, tgcg, tgcgc, tgcgcg\}. \end{aligned}$$

By checking the equivalence relations, 13 out of 18 elements in V are kept as representatives of the equivalent classes. Adding the class $[L'] \subset \Sigma^*$ we get all 14 equivalent classes of Σ^*:

$$[\epsilon]\ [g]\ [gc]\ [gcg]\ [gcgc]\ [gcgcg]\ [gcgcgc]\ [c]\ [cg]\ [cgc]\ [cgcg]\ [cgcgc]\ [cgcgcg]\ [L'].$$

At first glance the requirement of checking the equivalence relations for every $z \in \Sigma^*$ may seem formidable as it deals with an infinite set. However, a little practice shows that this may be done effectively without too much work.

Treating each class as a state, we define the discrete transfer function by

$$\delta([x_i], s) = [x_i s] \quad \text{for } x_i \in V \text{ and } s \in \Sigma.$$

The result is shown in Table 5.6. The special class $[L']$ is a "dead end", that is, an unacceptable state. The minDFA defined by the transfer function is drawn in Fig. 5.9.

Table 5.6 The transfer function for the minDFA accepting the *A. aeolicus* genome.

Class	a	c	g	t
$[\epsilon]$	$[\epsilon]$	$[c]$	$[g]$	$[c]$
$[g]$	$[\epsilon]$	$[gc]$	$[g]$	$[c]$
$[gc]$	$[\epsilon]$	$[c]$	$[gcg]$	$[c]$
$[gcg]$	$[\epsilon]$	$[gcgc]$	$[g]$	$[c]$
$[gcgc]$	$[\epsilon]$	$[c]$	$[gcgcg]$	$[c]$
$[gcgcg]$	$[\epsilon]$	$[gcgcgc]$	$[g]$	$[c]$
$[gcgcgc]$	$[L']$	$[c]$	$[L']$	$[c]$
$[c]$	$[\epsilon]$	$[c]$	$[cg]$	$[c]$
$[cg]$	$[\epsilon]$	$[cgc]$	$[g]$	$[c]$
$[cgc]$	$[\epsilon]$	$[c]$	$[cgcg]$	$[c]$
$[cgcg]$	$[\epsilon]$	$[cgcgc]$	$[g]$	$[c]$
$[cgcgc]$	$[\epsilon]$	$[c]$	$[cgcgcg]$	$[c]$
$[cgcgcg]$	$[\epsilon]$	$[L']$	$[g]$	$[c]$

Counting the number of lines leading from a node (state) to another, we write down the following *incidence* matrix:

$$M = \begin{pmatrix} 1 & 1 & 0 & 0 & 0 & 0 & 0 & 2 & 0 & 0 & 0 & 0 & 0 \\ 1 & 1 & 1 & 0 & 0 & 0 & 0 & 1 & 0 & 0 & 0 & 0 & 0 \\ 1 & 0 & 0 & 1 & 0 & 0 & 0 & 2 & 0 & 0 & 0 & 0 & 0 \\ 1 & 1 & 0 & 0 & 1 & 0 & 0 & 1 & 0 & 0 & 0 & 0 & 0 \\ 1 & 0 & 0 & 0 & 0 & 1 & 0 & 2 & 0 & 0 & 0 & 0 & 0 \\ 1 & 1 & 0 & 0 & 0 & 0 & 1 & 1 & 0 & 0 & 0 & 0 & 0 \\ 0 & 0 & 0 & 0 & 0 & 0 & 0 & 2 & 0 & 0 & 0 & 0 & 0 \\ 1 & 0 & 0 & 0 & 0 & 0 & 0 & 2 & 1 & 0 & 0 & 0 & 0 \\ 1 & 1 & 0 & 0 & 0 & 0 & 0 & 1 & 0 & 1 & 0 & 0 & 0 \\ 1 & 0 & 0 & 0 & 0 & 0 & 0 & 2 & 0 & 0 & 1 & 0 & 0 \\ 1 & 1 & 0 & 0 & 0 & 0 & 0 & 1 & 0 & 0 & 0 & 1 & 0 \\ 1 & 0 & 0 & 0 & 0 & 0 & 0 & 2 & 0 & 0 & 0 & 0 & 1 \\ 1 & 1 & 0 & 0 & 0 & 0 & 0 & 1 & 0 & 0 & 0 & 0 & 0 \end{pmatrix} \tag{5.15}$$

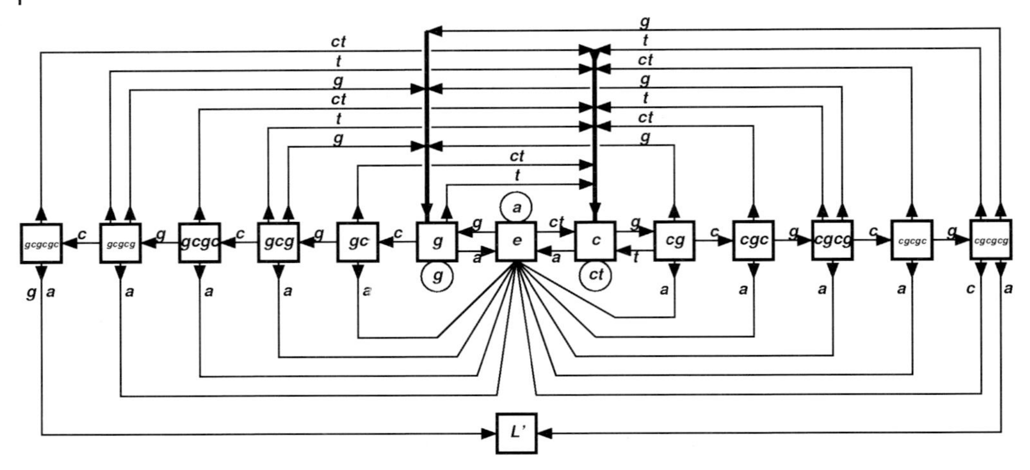

Fig. 5.9 A minDFA accepting the *A. aeolicus* genome with the four forbidden 7-strings given in Eq. (5.13).

To make a connection with the generating function (5.14) we note that the characteristic polynomial of M is related to $f(\frac{1}{\lambda})$:

$$\det(\lambda I - M) = \lambda^{13} f\left(\frac{1}{\lambda}\right).$$

Moreover, the sum of elements in the first row of the K-th power of M is nothing but a_K [9]:

$$a_K = \sum_{j=1}^{13} (M^K)_{1j}.$$

In order not to cause any confusion we note that although many true avoided strings longer than seven may be identified in the *A. aeolicus* genome, we have only used the $K = 7$ ones to construct the minDFA shown in Fig. 5.9. In fact, this DFA accepts a greater language of which $L(G)$ is a subset. In principle, one may invoke more forbidden words to construct a more complex DFA that accepts a smaller language still having $L(G)$ as a subset. Since our goal consists of calculating the number of redundant avoided strings of longer length caused by a given set of true avoided strings, it is enough to restrict the forbidden set to the true avoided strings up to a certain length.

5.6 Decomposition and Reconstruction of Protein Sequences

In recent years we have developed a composition vector tree (CVTree) approach [50–53] to infer phylogenetic relationships of bacteria from their complete genomes without making sequence alignment. The justification of the CVTree method has led to, among other things, the problem of uniqueness of reconstruction of a protein sequence from its constituent K-peptides. This problem has a natural relation to the number of Eulerian loops in a graph, a well-developed chapter of graph theory (see, for example, [54]). It turns out that in order to tell whether a given sequence has a unique reconstruction at a fixed K the notion of factorizable languages again comes to our help.

5.6.1 A Few Notions of Graph Theory

We need a few notions from graph theory. Look at a connected directed graph with a certain number of labeled nodes. If node i is connected to node j by one directed arc, then we say $a_{ij} = 1$, and so forth. From a beginning node v_b we go through a number of arcs to an ending node v_f in such a way that each arc has been traversed once and only once; such a path is called an *Eulerian path*. If $v_b = v_f$ the path becomes an *Eulerian loop*. A graph in which a Eulerian loop exists is called a *Euler graph*. A Eulerian path may be transformed into a Eulerian loop by drawing an auxiliary arc from v_f back to v_b.

From a given node, there may be d_{out} arcs going out to other nodes; d_{out} is called the outdegree of the node. Likewise, there may be d_{in} arcs coming into a node, d_{in} defines the indegree of the node. The condition for an undirected graph to be Eulerian was indicated by Euler in 1736, the year that has been considered the beginning of graph theory. In our case of a directed graph it may be formulated as

$$d_{\text{in}}(i) = d_{\text{out}}(i) = d_i$$

for all nodes numbered in a certain way from $i = 1$ to m. The numbers d_i are simply called degrees. We define a diagonal matrix

$$M = \text{diag}(d_1, d_2, \cdots, d_m).$$

The connectivity of the nodes is described by an adjacent matrix $A = \{a_{ij}\}$, where a_{ij} is the number of arcs leading from node i to j. From the matrices M and A we form the Kirchhoff matrix

$$C = M - A.$$

The Kirchhoff matrix has the peculiar property that its elements along any row or column sum to zero. Furthermore, for any $m \times m$ Kirchhoff matrix all $(m-1) \times (m-1)$ minors are equal and this common minor is denoted by Δ.

A graph is called *simple* if (1) there are no parallel arcs between nodes, that is, $a_{ij} = 0$ or $1\ \forall i, j$; and (2) there are no rings at any node, that is, $a_{ii} = 0\ \forall i$.

The number R of Eulerian loops in a simple Euler graph is given by the **BEST formula** (BEST stands for N. G. de **B**ruijn, T. van Aardenne-**E**hrenfest, C. A. B. **S**mith, and W. T. **T**utte) [54]:

$$R = \Delta \prod_i (d_i - 1)!.$$

This formula gives the number of Eulerian loops in a Euler graph without specifying a starting node. If a node k is specified as the beginning (hence ending) of the loop, then the number of loops starting from k is [55]

$$R = \Delta d_k \prod_i (d_i - 1)!, \tag{5.16}$$

where d_k is the degree of the node k.

In a general Euler graph there may be parallel arcs between certain pairs of nodes ($a_{ij} > 1$) and rings at some nodes ($a_{ii} \neq 0$). One may put auxiliary nodes on these arcs and rings to make the graph simple. By applying elementary operations to the larger Kirchhoff matrix thus obtained, one can reduce it to the original size with some $a_{ii} \neq 0$ and $a_{ij} > 1$. Since the parallel arcs and rings are unlabeled we must eliminate the redundancy in the counting result. Therefore, the BEST formula is modified to

$$R = \frac{\Delta d_k \prod_i (d_i - 1)!}{\prod_{ij} a_{ij}!}. \tag{5.17}$$

As $0! = 1! = 1$ this formula reduces to the previous one in case of simple graphs. The modified BEST formula (5.17) first appeared in [56] where Eulerian loops from a fixed starting node were considered.

5.6.2
The *K*-Peptide Representation of Proteins

In the CVTree method, instead of using a primary protein sequence made of M amino acid letters, we decompose the protein into $M - K + 1$ overlapping K-peptides. Take, for example, the human centromere protein B (*CENB_HUMAN* in the SwissProt database), consisting of 599 amino acids:

```
MGPKRRQLTF REKSRIIQEV EENPDLRKGE IARRFNIPPS TLSTILKNKR AILASERKYG
VASTCRKTNK LSPYDKLEGL LIAWFQQIRA AGLPVKGIIL KEKALRIAEE LGMDDFTASN
GWLDRFRRRH GVVSCSGVAR ARARNAAPRT PAAPASPAAV PSEGSGGSTT GWRAREEQPP
SVAEGYASQD VFSATETSLW YDFLPDQAAG LCGGDGRPRQ ATQRLSVLLC ANADGSEKLP
PLVAGKSAKP RAGQAGLPCD YTANSKGGVT TQALAKYLKA LDTRMAAESR RVLLLAGRLA
AQSLDTSGLR HVQLAFFPPG TVHPLERGVV QQVKGHYRQA MLLKAMAALE GQDPSGLQLG
LTEALHFVAA AWQAVEPSDI AACFREAGFG GGPNATITTS LKSEGEEEEE EEEEEEEEEG
EGEEEEEEGE EEEEEGGEGE ELGEEEEVEE EGDVDSDEEE EEDEESSSEG LEAEDWAQGV
VEAGGSFGAY GAQEEAQCPT LHFLEGGEDS DSDSEEEDDE EEDDEDEDDD DDEEDGDEVP
VPSFGEAMAY FAMVKRYLTS FPIDDRVQSH ILHLEHDLVH VTRKNHARQA GVRGLGHQS
```

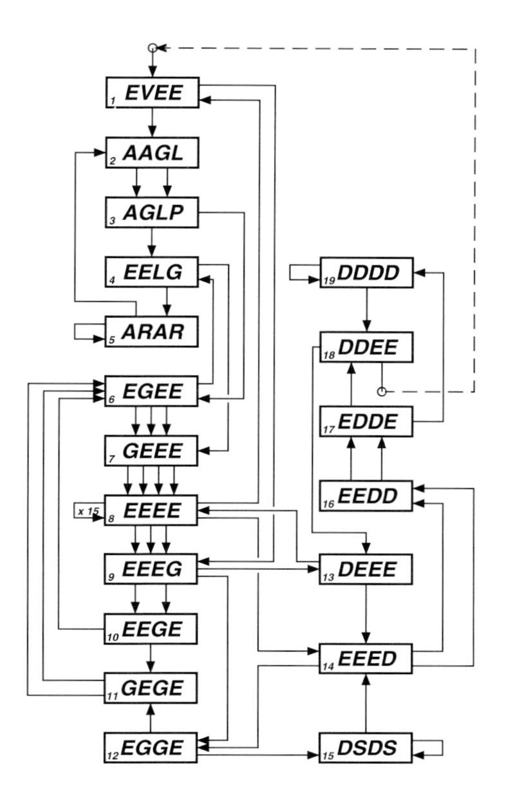

Fig. 5.10 A Euler graph generated by the protein sequence CENB_HUMAN with 599 amino acids.

Decomposing the above sequence into a collection of overlapping 5-peptides MGPKR, GPKRR, and so forth., and considering each 5-peptide as a transition from its 4-letter prefix to the 4-letter suffix, one easily transforms the sequence to a Eulerian path, using the 4-strings as node labels. Since we are interested in the number of Eulerian loops, we can replace a node with degree 1 by an arc without affecting the number of loops. Thus, only nodes with $d_i > 1$ matter. Finally, by drawing an auxiliary arc from the last node to the first we get a Eulerian loop and this loop defines an Euler graph, see Fig. 5.10. Since we are interested only in the number of different Eulerian loops in the graph, we can add a degree 1 node v_0 on the auxiliary arc and count the number of loops starting from this node. Therefore, in practical calculations we always take $d_k = 1$ in Eqs. (5.16) and (5.17).

From Fig. 5.10 we infer the diagonal matrix

$$M = \mathrm{diag}(1,2,2,2,2,2,4,4,20,4,2,2,2,2,3,2,2,2,2,2)$$

and the adjacent matrix

$$A = \begin{pmatrix}
0 & 1 & 0 & 0 & 0 & 0 & 0 & 0 & 0 & 0 & 0 & 0 & 0 & 0 & 0 & 0 & 0 & 0 & 0 & 0 \\
0 & 0 & 1 & 0 & 0 & 0 & 0 & 0 & 0 & 1 & 0 & 0 & 0 & 0 & 0 & 0 & 0 & 0 & 0 & 0 \\
0 & 0 & 0 & 2 & 0 & 0 & 0 & 0 & 0 & 0 & 0 & 0 & 0 & 0 & 0 & 0 & 0 & 0 & 0 & 0 \\
0 & 0 & 0 & 0 & 1 & 0 & 1 & 0 & 0 & 0 & 0 & 0 & 0 & 0 & 0 & 0 & 0 & 0 & 0 & 0 \\
0 & 0 & 0 & 0 & 0 & 1 & 0 & 1 & 0 & 0 & 0 & 0 & 0 & 0 & 0 & 0 & 0 & 0 & 0 & 0 \\
0 & 0 & 1 & 0 & 0 & 1 & 0 & 0 & 0 & 0 & 0 & 0 & 0 & 0 & 0 & 0 & 0 & 0 & 0 & 0 \\
0 & 0 & 0 & 0 & 1 & 0 & 0 & 3 & 0 & 0 & 0 & 0 & 0 & 0 & 0 & 0 & 0 & 0 & 0 & 0 \\
0 & 0 & 0 & 0 & 0 & 0 & 0 & 0 & 4 & 0 & 0 & 0 & 0 & 0 & 0 & 0 & 0 & 0 & 0 & 0 \\
0 & 1 & 0 & 0 & 0 & 0 & 0 & 0 & 15 & 3 & 0 & 0 & 0 & 0 & 1 & 0 & 0 & 0 & 0 & 0 \\
0 & 0 & 0 & 0 & 0 & 0 & 0 & 0 & 0 & 0 & 2 & 0 & 1 & 1 & 0 & 0 & 0 & 0 & 0 & 0 \\
0 & 0 & 0 & 0 & 0 & 0 & 1 & 0 & 0 & 0 & 0 & 1 & 0 & 0 & 0 & 0 & 0 & 0 & 0 & 0 \\
0 & 0 & 0 & 0 & 0 & 0 & 2 & 0 & 0 & 0 & 0 & 0 & 0 & 0 & 0 & 0 & 0 & 0 & 0 & 0 \\
0 & 0 & 0 & 0 & 0 & 0 & 0 & 0 & 0 & 0 & 0 & 1 & 0 & 0 & 0 & 1 & 0 & 0 & 0 & 0 \\
0 & 0 & 0 & 0 & 0 & 0 & 0 & 0 & 1 & 0 & 0 & 0 & 0 & 0 & 1 & 0 & 0 & 0 & 0 & 0 \\
0 & 0 & 0 & 0 & 0 & 0 & 0 & 0 & 0 & 0 & 0 & 0 & 1 & 0 & 0 & 0 & 2 & 0 & 0 & 0 \\
0 & 0 & 0 & 0 & 0 & 0 & 0 & 0 & 0 & 0 & 0 & 0 & 0 & 0 & 1 & 1 & 0 & 0 & 0 & 0 \\
0 & 0 & 0 & 0 & 0 & 0 & 0 & 0 & 0 & 0 & 0 & 0 & 0 & 0 & 0 & 0 & 0 & 2 & 0 & 0 \\
0 & 0 & 0 & 0 & 0 & 0 & 0 & 0 & 0 & 0 & 0 & 0 & 0 & 0 & 0 & 0 & 0 & 0 & 1 & 1 \\
1 & 0 & 0 & 0 & 0 & 0 & 0 & 0 & 0 & 0 & 0 & 0 & 0 & 1 & 0 & 0 & 0 & 0 & 0 & 0 \\
0 & 0 & 0 & 0 & 0 & 0 & 0 & 0 & 0 & 0 & 0 & 0 & 0 & 0 & 0 & 0 & 0 & 0 & 1 & 1
\end{pmatrix} \quad (5.18)$$

The common minor of the Kirchhoff matrix $C = M - A$ is $\Delta = 168\,960$. Denoting by $R(K)$ the number of reconstructions at K, we get from the modified BEST formula $R(5) = 491\,166\,720$. We see that the protein *CENB_HUMAN* belongs to the few sequences that have a huge number of reconstructions at moderate value of K. Most of the naturally occurring proteins have a unique reconstruction at $K = 5 \sim 6$ [57], a fact speaking in favor of the composition vector approach.

In principle, one can write a program to generate all distinct reconstructions from a given set of K-tuples obtained by decomposing an original protein sequence. In so doing a cut-off reconstruction number must be set because there might be proteins with a huge number of reconstructions as shown by the above example. Alternatively, by implementing the modified BEST formula (5.17) one may get the number of reconstructions without producing all the sequences. It is curious to note that equipped with these programs one can fish out a set of proteins that have a huge number of reconstructions without any biological knowledge as a prerequisite [58].

However, if we are only interested in whether a given symbolic sequence over a finite alphabet Σ has a unique reconstruction at a given K or not, factorizable language may help us to construct finite state automata to yield an YES/NO answer.

5.6.3 Uniquely Reconstructible Sequences and Factorizable Language

We put the sequence decomposition and reconstruction problem in a more general setting and ask a more specific question. Consider a finite alphabet Σ and an arbitrary sequence $s \in \Sigma^*$ of length N. Decompose the sequence s into $N - K + 1$ overlapping K-tuples and then reconstruct sequences by using each of the K-tuples once and only once. Collect all the sequences that have a unique reconstruction into a subset $L \subset \Sigma^*$. The subset L defines the language of uniquely reconstructible sequences. The language L is a factorizable language by definition, because any substring in a word $s \in L$ must be uniquely reconstructible otherwise the whole word s cannot be in L. To the contrary, the language $L' = \Sigma^* - L$ consists of sequences that must have two or more reconstructions. Among words in L' there is a set L'' of forbidden words.

Question: Given a sequence $s \in \Sigma^*$, judge whether s is in L or not; if not, s must be in L'. A deterministic finite state automaton may be built to answer this YES/NO question.

It was first conjectured in [59] and then proved in [60] that the non-uniqueness in sequences reconstruction comes from two kinds of transformations. Applied to our problem of loops with starting and ending nodes connected by an auxiliary arc, we can take all the $(K - 1)$-tuples that label the nodes as symbols in a new alphabet. Then it is sufficient to consider only the case of $K = 2$. The two types of transformations are: *Transposition*

$$\cdots xwz \cdots xuz \cdots \Longleftrightarrow \cdots xuz \cdots xwz \cdots ,$$
$$\cdots xwxux \cdots \Longleftrightarrow \cdots xuxwx \cdots ,$$

and *Rotation*

$$xwzux \Longleftrightarrow zuxwz,$$

where $x, z \in \Sigma$ and $w, u \in \Sigma^*$. These conditions are necessary but not sufficient for the non-uniqueness of reconstruction. Since the starting and ending symbols are fixed in our construction the *rotation* drops out of consideration.

A recent paper [61] discussed the reconstruction problem in an entirely different context and obtained the necessary and sufficient conditions for a Eulerian loop to be unique in a Euler graph. In particular, Kontorovich [61] proved

Theorem 5.6 *The set of all uniquely reconstructible sequences is a regular language.*

Unfortunately, the abstract proof of the theorem did not provide a way to build a finite state automaton that would accept a uniquely reconstructible language.

By invoking the notion of factorizable language and using the set of forbidden words we construct such an automaton explicitly and thus provide a constructive proof of the theorem.

We begin with

Theorem 5.7 *Under the condition that $K = 2$ and the starting and ending symbols are identical, a uniquely reconstructible language $L \subset \Sigma^*$ only possesses two types of forbidden words:*

$$\begin{array}{ll} (i) & x\alpha y\gamma x\beta y, \\ (ii) & x\alpha x\beta x, \end{array} \tag{5.19}$$

where $x, y \in \Sigma$ and $x \neq y$, $\alpha, \beta, \gamma \in L$, and they satisfy the following conditions

1. *In (i): at least one of α and β is not empty, all α, β and γ do not contain neither the symbols x, y nor identical symbols.*
2. *In (ii): at least one of α and β is not empty, they do not contain neither the symbol x nor identical symbols.*

We skip the simple but somewhat lengthy proof and go directly to the construction of the finite state automaton.

5.6.4
Automaton that Accepts a Uniquely Reconstructible Language

We first describe a deterministic finite state automaton M that accepts a uniquely reconstructible language L [62]. The automaton M consists of five elements:

$$M = \{Q, \Sigma, \delta, q_0, F\}, \tag{5.20}$$

where Σ is the finite alphabet, $Q = \{q\}$ is the set of states, $q_0 \in Q$ is the initial state, $F \in Q$ are the states that accept L, and δ is the transfer function from $Q \times \Sigma$ to Q. The automaton M reads in symbols in a sequence $s \in \Sigma^*$ from the left to right, one symbol a at a time, and changes its state from q to a new state q' according to the transfer function $\delta(q, a) = q'$. We explain these elements one by one.

1. Σ is an alphabet of m symbols $\{a_1, a_2, \cdots, a_m\}$ which may be denoted by $\{1, 2, \cdots, m\}$ as well. The language $L \subset \Sigma^*$ is defined over Σ and the automaton M reads a sequence $s \in \Sigma^*$.
2. Each state in Q consists of three components $(p; n; c)$: $Q = P \times N \times C = \{(p; n; c)\}$, where

- p records the most recently read symbol $a_p \in \Sigma$. For the initial state q_0 when no input symbol has been read yet we introduce a symbol a_0 (or 0) which does not belong to Σ. Thus we may write $P = \Sigma \bigcup 0$.
- n is a list of $m+1$ symbols $(n_0, n_1, n_2, \cdots, n_m)$ that updates the next symbol read after p. For the initial state $n = (\epsilon, \epsilon, \epsilon, \cdots, \epsilon) \equiv \epsilon^{m+1}$, where ϵ means empty or non-existence. Thus $N = (\Sigma \bigcup \epsilon)^{m+1}$.
- c is a list of m toggle switches: $c = (c_1, \cdots, c_m)$. We denote the two states of a toggle as $WHITE$ and $BLACK$. Initially $c = WHITE^m$. Whenever a forbidden word in the sequence s has been read c becomes $BLACK^m$. As long as c is not all-black the state q is acceptable. Once c becomes all-black it remains so for ever and the automaton recognizes s as a non-uniquely reconstructible sequence.

3. The initial state $q_0 = (0; \epsilon^{m+1}; WHITE^m)$.
4. The acceptable states

$$F = (p; n; c \neq BLACK^m). \tag{5.21}$$

5. The key element of M is the transfer function $\delta(q, a)$. A program to implement δ is written down below using a simple meta-language (i is a working variable in the program):

```
1   procedure δ((p, n, c), a)
2     if (n_p ≠ ε) & (n_p ≠ a) then
3       i ← p
4       repeat
5         c_i ← BLACK
6         i ← n_i
7       until i = p
8     endif
9     if c_a = BLACK then
10      c ← BLACK^m
11    endif
12    p ← n_p ← a
13  end procedure
```

All technical terms involved in describing the automaton may be found in [4]. A C++ implementation of the transfer function δ is available upon request to the authors of [62]. As usual, the best way to understand the workings of this program is to take a couple of uniquely and non-uniquely reconstructible sequences over some alphabet and follow the transitions among states according to the program.

The proof of the following theorem may help to grasp the essence of the transfer function δ.

Theorem 5.8 *The language $L(M)$ accepted by the finite state automaton M defined in Eq. (5.20) is the uniquely reconstructible language $L \subset \Sigma^*$ over an alphabet Σ of m symbols.*

Proof. The proof of $L(M) = L$ goes in two parts.

First part. We prove $L(M) \subset L$ by showing that all sequences in the complementary language L' cannot be accepted by the automaton M.

Suppose $t \notin L$ then t contains forbidden words. Consider the first forbidden word encountered when feeding t to M. If this forbidden word belongs to type (i) in Theorem 5.7, that is, it is of type $x\alpha y\gamma x\beta y$, where $x, y \in \Sigma$, $x \neq y$, $\alpha, \beta, \gamma \in L$, α and β are not empty at the same time, α, β and γ do not contain neither x, y nor identical symbols. Now look at Lines 2 $\sim$ 8 in procedure δ. Suppose that M has reached a state where p equals the second x, then its next symbol is the a in $\delta(q, a)$. (If β is not empty then a is the first symbol in β, otherwise a is the last y.) Since the symbol after the first x cannot be a, the condition in Line 2 of the procedure holds. Therefore, Line 5 makes $c_x = BLACK$. Now the loop from Line 4 to Line 7 makes every toggle c_i corresponding to symbols in $x\alpha y\gamma$ becomes $BLACK$, including $c_y = BLACK$. As blackened toggles cannot become $WHITE$ again, when reading in the last y the execution of Line 9 and 10 turns c to all-$BLACK$, that is, M enters a non-acceptable state.

When the first forbidden word in t belongs to type (ii) in Theorem 5.7, that is, it is of type $x\alpha x\beta x$, the situation is simpler and we ignore the discussion.

Second part. In order to prove $L \subset L(M)$ it is enough to show that after reading in the sequence $t \in L$ the state $(p; n; c)$ belongs to the acceptable states F given in Eq. (5.21), namely, the toggles c are not all-$BLACK$.

We prove this by mathematical induction with respect to the length $|t| = N$ of the sequence $t \in L$. From the definition of the automaton M we know that when $N = 0$ the initial state $q_0 \in F$, the statement holds true. Now suppose the statement holds for $N - 1$ and discuss the situation of $|t| = N$.

Since L is a factorizable language, we denote by a the last symbol of $t \in L$ and write t as $t = sa$, the prefix s of t is uniquely reconstructible and it is of length $N - 1$. Therefore, c is not all-$BLACK$ after reading in s.

Now we prove that upon reading a the toggles c do not become all-$BLACK$ by reduction to absurdity. Suppose the opposite is true, that is, upon reading in a the toggles c become all $BLACK$. An inspection of the procedure δ shows that this may happen in two cases:

(1) The toggle c_a has become $BLACK$ before reading in a. Therefore, after reading a the Line 10 in procedure δ is executed that makes c all-$BLACK$. However, in order to enable Line 4 to be executed the two conditions in Line 1 must be satisfied. This means the sequence s must have a suffix like $b\alpha b\beta$ where the symbols following the two symbols b must be different and at the same time $a \in b\alpha$. Then $t = sa$ must contain a forbidden word as suffix no matter whether a equals b or not. Therefore, $t \notin L$, a contradiction.

(2) The toggle c_a was $WHITE$ and it turns $BLACK$ only after reading a. Now we must consider the last symbol in s. Denote it by p. The fact that changing c_a to $BLACK$ takes place in Line 2 to Line 7 shows that t must have a suffix $p\alpha pa$ with non-empty α; the first symbol of α is not a but a appears in $p\alpha$. Therefore, $p\alpha pa$ is a forbidden word independent on whether p equals a or not, a contradiction to $t \in L$.

Thus we have proved $L \subset L(M)$. Combination of the two parts completes the proof. □

Clearly, M is a deterministic finite state automaton. Although how to build the corresponding minDFA from the above DFA is known in principle [4], the explicit construction remains lacking for the time being. In particular, we do not know the size of the minDFA which is given by the index of the equivalence relation R_L generated by L in Σ^*.

In principle, one can build an automaton that accepts the complementary L' of the language L. The right-linear grammar to implement such an NDFA may be found in [62].

5.6.5 Other Applications of the Sequence Reconstruction Problem

The sequence unique reconstruction problem occurred in so-called *sequencing by hybridization* (SBH), one of the earliest proposed application of DNA arrays. It has been analyzed from the viewpoint of the number of Eulerian loops (see [60] and references therein). However, no connection with language theory was mentioned in these studies.

Another possible application is associated with sequence randomization under constraints. In order to tell the statistical significance of certain "signals" revealed in some symbolic sequences one must compare them with what might be observed in a background model [44]. A frequent choice of the background model is full randomization of the original sequence that only keeps the number of single letters unchanged. However, under certain circumstances it is more appropriate to perform the randomization under some further constraints, for example, keeping a designated number of short strings fixed and having the rest of the sequence "randomized". In this setting we encounter the opposite of the unique reconstruction problem, namely, only when there exists a huge number of reconstructions under the given constraints does it make sense to speak about "randomization", otherwise it is just a choice among a finite number of shufflings. This problem has been studied before [63] and again no reference to factorizable language was made.

We expect more applications of factorizable languages, especially the method of forbidden words, in various problems of dynamics and biology.

References

1 Hao, B.-L., in *On Self-Organization – An Interdisciplinary Search for a Unifying Principle*, Springer Series in Synergetics 61, Springer Verlag, **1994**, p. 197

2 Hao, B.-L., Zheng, W.-M., *Applied Symbolic Dynamics*, World Scientific, Singapore, **1998**

3 Shannon, C. E., A mathematical theory of communication, *Bell Systems Tech. J.*, 27 (**1948**), p. 379 and p. 623

4 Hopcroft, J and Ullman, J., *Introduction to Automata Theory, Languages and Computation*, Addison-Wesley, Reading, **1979**

5 Xie, H.-M., *Grammatical Complexity and One-Dimensional Dynamical Systems*, World Scientific, Singapore, **1996**

6 Rozenberg, G., Salomma, A., eds., *Handbook of Formal Languages*, Vols. 1-3, Springer, **1997**

7 Shyr, H. J., *Free Monoids and Languages*, Hon Min Book Company, Taichung, **1991**

8 Rozenberg, G., Salomaa, A., *The Mathematical Theory of L-Systems*, Academic Press, **1980**

9 Wolfram, S., Computation theory of cellular automata. *Commun. Math. Phys.* 96 (**1984**), p. 15

10 Morse, M., Hedlund, G. A., Symbolic dynamics, *Am. J. Math.* 60 (**1938**), p. 815; reprinted in *Collected Papers of M. Morse*, vol. 2, World Scientific, **1986**

11 Hao, B.-L., Xie, H.-M., Yu, Z.-G., Chen, G.-Y., Factorisable language: From dynamics to complete genomes, *Physica* A288 (**2000**) p. 10

12 Xie, H.-M., On formal languages of one-dimensional dynamical systems, *Nonlinearity*, 6, (**1993**), p. 997

13 Hao, B.-L., Symbolic dynamics and characterization of complexity, *Physica* D51 (**1991**), p. 161

14 Crutchfield, J. P. and Young, K., Computation at the onset of chaos, in *Complexity, Entropy, and Physics of Information*, ed. W. Zurek, Addison-Wesley, **1990**, p. 223

15 Wang, Y., Yang, L., Xie, H.-M., Complexity of unimodal maps with aperiodic kneading sequences, *Nonlinearity*, 12 (**1999**), p. 1151

16 von Neumann, J., *Theory of Self-Reproducing Automata*, edited by Burks, A. W., Univ of Illinois Press, Champaign, **1966**

17 Berlekamp, E., Conway, J. H. and Guy, R., *Winning Ways*, vol 2, Academic Press, New York, **1982**

18 Wolfram, S., *Theory and Applications of Cellular Automata*, Singapore: World Scientific, **1986**

19 Wolfram, S., *Cellular Automata and Complexity*, Addison-Wesley, New York, **1994**

20 Wolfram, S., *A New Kind of Science*, Wolfram Media Inc, Champaign, **2002**

21 Cook, M., Universality of elementary cellular automata. *Complex Systems* 15 (**2004**), p. 1

22 Hurd, L. P., Recursive cellular automata invariant sets. *Complex Systems* 4 (**1990**), p. 119

23 Hurd, L. P., Nonrecursive cellular automata invariant sets. *Complex Systems* 4 (**1990**), p. 131

24 Culik, II K., Hurd, L. P. and Yu, S., Computation theoretic aspects of cellular automata. *Physica* D45 (**1990**), p. 357

25 Kari, J., The nilpotency problem of one-dimensional cellular automata. *SIAM J Comp* 21 (**1992**), p. 571

26 Xie, H. M., Distinct excluded blocks and grammatical complexity of dynamical systems, *Complex Systems* 9 (**1995**), p. 73

27 Xie, H.-M., The complexity of limit languages of cellular automaton: an example. *J. of Systems Science & Complexity* 14 (**2001**), p. 17

28 Jiang, Z.-S. and Wang, Y., Complexity of limit language of the elementary cellular automaton of rule 22. *Appl. Math. J. Chinese Univ, ser. B* 20 (**2005**), p. 268

29 Jiang, Z.-S., A complexity analysis of the elementary cellular automaton of rule 122. *Chinese Science Bulletin* 46(7) (**2001**), p. 600

30 Cattaneo, G., Dennunzio, A. and Margara, L., Chaotic subshifts and related languages – applications to one-dimensional cellular automata. *Fund Inform* 52 (**2002**), p. 39

31 Hurley, M., Attractors in restricted cellular automata. *Proc. Amer. Math. Soc.* 115 (**1990**), p. 563

32 Blanchard, F., Kůrka, P. and Maass, A., Topological and measure-theoretic properties of one-dimensional cellular automata. *Physica* D103 (**1997**), p. 86

33 Kůrka, P., Languages, equicontinuity and attractors in cellular automata. *Ergodic Theory & Dynamic Systems* 17 (**1997**), p. 417

34 Jen, E., Aperiodicity in one-dimensional cellular automata. *Physica* D45 (**1990**), p. 9

35 Jen, E., Exact solvability and quasiperiodicity of one-dimensional cellular automata. *Nonlinearity* 4 (**1990**), p. 251

36 Jiang, Z.-S. and Xie, H.-M., Evolution complexity of the elementary cellular automaton rule 18. *Complex Systems* 13 (**2001**), p. 271

37 Wang, Y. and Kenichi, M., Complexity of evolution languages of the elementary cellular automaton of rule 146. *Appl Math J Chinese Univ, Ser B* 21 (**2006**), p. 418

38 Wang, Y. and Jiang, Z.-S., Evolution complexity of the elementary cellular automaton of rule 22. *Appl Math J Chinese Univ, Ser B* 17 (**2002**), p. 404

39 Qin, D.-K. and Xie, H.-M., Complexity analysis of time series generated by elementary cellular automata. *App Math J Chinese Univ, Series B* 20 (**2005**), p. 253

40 Qin, D.-K. and Xie, H.-M., Catalan numbers, Dyck language and time series of elementary cellular automaton of rule 56. *J. of Systems Science & Complexity* 18 (**2005**), p. 404

41 Sloane, N. J., Sequence A000108 in *The On-Line Version of the Encyclopedia of Integer Sequences*: http: //www.research.att.com/~njas/sequences

42 Shen, J.-J., Zhang, S.-Y., Lee, H.-C., Hao, B.-L., SeeDNA: Visualization of K-string content of long DNA sequences and their randomized counterparts, *Genomics, Proteomics & Bioinformatics* 2 (**2004**), p. 192

43 Hao, B.-L., Lee, H.-C., Zhang, S.-U., Fractals related to long DNA sequences and bacterial complete genomes, *Chaos, Solitons & Fractals,* 11 (**2000**) p. 825

44 Xie, H.-M., Hao, B.-L. Visualization of *k*-tuples distribution in prokaryote complete genomes and their randomized counterparts, in *Bioinformatics. CBS2002 Proceedings,* , IEEE Computer Society, Los Alamitos, California **2002**, p. 31

45 Deckert,G. et al., The complete genome of the hyperthermophilic bacterium *Aquifex aeolicus*, *Nature* 392 (**1998**), p. 353

46 Hao, B.-L., Xie, H.-M., Yu, Z.-G., Chen, G.-Y., A combinatorical problem related to avoided strings in bacterial complete genomes, *Ann. Combin.* 4 (**2000**), p. 247

47 Hao, B.-L., Fractals from genomes: exact solutions of a biology-inspired problem, *Physica* A282 (**2000**) p. 225

48 Goulden, I., Jackson, D. M., *Combinatorial Enumeration*, John Wiley & Sons, New York, **1983**

49 Zhou, C., Xie, H.-M., Exact distribution of the occurrence number for K-tuples over an alphabet of non-equal probability letters, *Ann. Combinatorics* 8 (**2004**) p. 499

50 Qi, J., Wang, B., Hao, B.-L., Whole genome prokaryote phylogeny without sequence alignment: a K-string composition approach, *J. Mol. Evol.*, 58 (**2004**), p. 1

51 Qi, J., Luo, H., Hao, B.-L., CVTree: a phylogenetic tree reconstruction tool based on whole genomes, *Nucl. Acids Res.*, 32 (**2004**), Web Server Issue, p. W45

52 Hao, B.-L., Qi, J., Prokaryote phylogeny without sequence alignment: from avoidance signature to composition distance, *J. Bioinformatics & Computational Biology*, 2 (**2004**), p. 1

53 Gao, L., Qi, J., Sun, J.-D., Hao, B.-L., Prokaryote phylogeny meets taxonomy: an exhaustive comparison of the composition vector tree with the biologist's systematic bacteriology, *Science in China* Series C Life Sciences, to appear, (**2007**)

54 Fleischner, H., *Eulerian Graphs and Related Topics*, Part 1, vol. 2, **1991**, p. IX80

55 Bollobás, B., *Modern Graph Theory*, Springer-Verlag, New York, **1998**

56 Hutchinson, J. P., On words with prescribed overlapping subsequences, *Utilitas Mathematica*, 7, (**1975**), p. 241

57 Xia, L., Zhou, C., Phase transitions in sequence unique reconstruction, *J. Syst. Sci. & Complexity* 20 (**2007**), p. 18

58 Shi, X.-L., Xie, H.-M., Zhang, S.-Y., Hao, B.-L., Decomposition and reconstruction of protein sequences: the problem of uniqueness and factorizable language, *J. Korean Phys. Soc.* 58 (**2007**), p. 118

59 Ukkonen, E., Approximate string matching with q-grams and maximal matches, *Theor. Comput. Sci.*, 92, (**1992**), p. 191

60 Pevzner, P. A., *Computational Molecular Biology: An Algorithmic Approach*, The MIT Press, Cambridge, MA, **2000**

61 Kontorovich, L., Uniquely decodable n-gram embeddings, *Theor. Computer Sci.*, 329 (**2004**), p. 271

62 Li, Q., Xie, H.-M., Finite automata for testing uniqueness of Eulerian trails, **2005**, arXive: cs.CC/0507052 (20 July 2005); Finite automata for testing composition-based reconstructibility of sequences, submitted to *J. Computer & Systems Sci.*, in April **2007**

63 Kandel, D., Matias, Y., Unger, R., Winkler, P., Shuffling biological sequences, *Discrete Appl. Math.*, 71 (**1996**), p. 171

6
Controlling Collective Synchrony by Feedback

Michael Rosenblum and Arkady Pikovsky

6.1
What is Collective Synchrony?

Synchronization is a general nonlinear phenomenon which can be briefly described as an adjustment of rhythms of self-sustained oscillatory systems due to their interaction [1]. In the simplest setup, two periodic oscillators of this class can adjust their phases and frequencies, even if the coupling between the systems is very weak. This effect is often called phase locking or frequency entrainment. The concept can be generalized to the case of many oscillating objects with a variety of different coupling configurations: oscillating subsystems can be placed on a regular lattice, or organized in a complex, possibly irregular, network, or form a continuous medium. If an oscillating unit is interacting with its neighbors only, then synchronization typically appears in a form of waves or oscillatory modes. Contrary to this, if the interaction is a long-range one, then global synchrony, where all or almost all units oscillate in pace, can set in. Examples of synchronous dynamics range from mechanical systems like pendulum clocks [2,3] and metronomes [4], through modern physical devices, e.g., Josephson junctions and lasers (see [5,6] and references therein) to live systems (see, e.g., a review [7]), including human beings [8,9].

The mostly popular model, describing collective dynamics in a large population of self-sustained oscillators is that of globally (all-to-all) coupled units. This framework has been used to describe many physical, biological, and social phenomena. The main effect observed in these models is the *collective synchrony*, when a large part or all units adjust their rhythms and produce a nonzero mean field, which has the same frequency as the synchronized majority. In the simplest setup this state appears from the fully asynchronous one via the *Kuramoto transition* [10,11]: if the coupling strength ε in the ensemble exceeds some threshold value ε_{cr}, the macroscopic mean field appears and its amplitude growth with the super-criticality $\varepsilon - \varepsilon_{cr}$. This transition is often considered in an analogy to second order phase transitions; on the other hand, it can be viewed at as a supercritical Hopf bifurcation for the mean field.

Review of Nonlinear Dynamics and Complexity. Edited by Heinz Georg Schuster

ISBN: 978-3-527-40729-3

Complexity of individual oscillators and/or of the coupling function can essentially complicate the behavior of the ensemble and result in interesting dynamics like chaos of the mean field, clustering and multistability, splay states, etc. [12–17]. Quite involved dynamics, when synchronous clusters have different frequencies, thus leading to quasiperiodicity, can appear in non-homogeneous populations [18]. Furthermore, quasiperiodic dynamics, when oscillators do not form any clusters and *are not locked* to the periodic mean field they produce, but remain, however, coherent, can be observed in homogeneous populations. Such counter-intuitive *partially synchronous* states have been first observed by van Vreeswijk in a model of delay-coupled integrate-and-fire neurons [19], see also [20], and later in [21]. As has been shown recently, such regimes naturally appear in case of nonlinear coupling between the units [22].

A transition to collective synchrony has been also observed for chaotic oscillating units [23–25]. Here the mechanism of synchrony is different from that in periodic oscillators: a non-zero mean field cannot completely adjust the individual rhythms, but just imposes a regular component in them. The global feedback due to coupling for this regular component suffices to maintain a periodic collective dynamics, while the regimes of individual units remain chaotic. For a large number of interacting chaotic oscillators one can average over individual chaos, so that the transition to synchrony in the mean field may be described, similar to the case of Kuramoto transition, as a Hopf bifurcation (although other scenarios, including subcritical bifurcations, are also possible).

Kuramoto-like models of all-to-all coupled units have been widely used to describe collective neuronal dynamics; a useful review can be found in [26]. Although analytical treatment is possible only for fully connected ensembles (what is certainly not true in real systems), this approximation is considered to be reasonable, due to a relatively high connectivity in neuronal networks. We illustrate the appearance of the collective synchrony in an ensemble of irregularly bursting neurons in Fig. 6.1. For a model of individual units we take a computationally efficient discrete model, the Rulkov map [27,28]. The equations of the ensemble read:

$$\begin{aligned} x_i(n+1) &= \frac{4.3}{1+x_i^2(n)} + y_i(n) + \varepsilon X(n) , \\ y_i(n+1) &= y_i(n) - 0.01(x_i(n)+1) , \end{aligned} \tag{6.1}$$

where n is the discrete time, $i = 1, \ldots N$ is the index of a neuron in the population, and

$$X(n) = \frac{1}{N_l} \sum_{1}^{N} x_i(n)$$

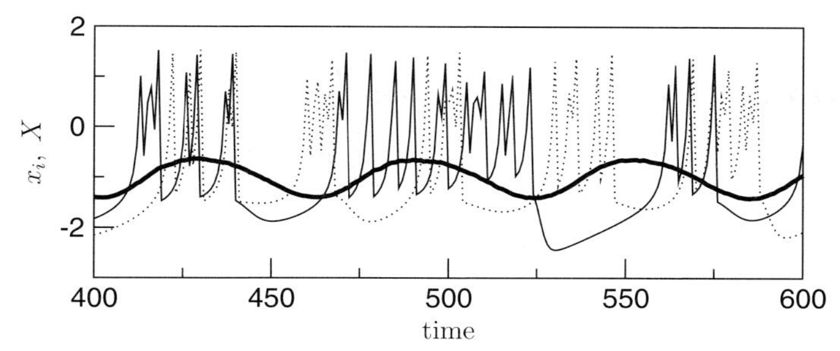

Fig. 6.1 Two bursting chaotic Rulkov neurons (solid and dotted lines), randomly picked up from a population, and the mean field of the population (bold line). It can be seen that there is some tendency to synchronization between neurons on the low time scale, i.e. in bursting, and no synchrony in spiking. The collective mode has a very pronounced regular component.

is the field, acting on a neuron. Each neuron generates chaotic bursts, and when the coupling ε exceeds the critical value $\varepsilon_{cr} \approx 0.055$, the mean field starts oscillating, see the upper curve in Fig. 6.2. This oscillations, as can be seen from Fig. 6.1, are quite regular.

Using this example we illustrate in Fig. 6.2 that the assumption of all-to-all coupling gives a good approximation for not globally, but randomly coupled

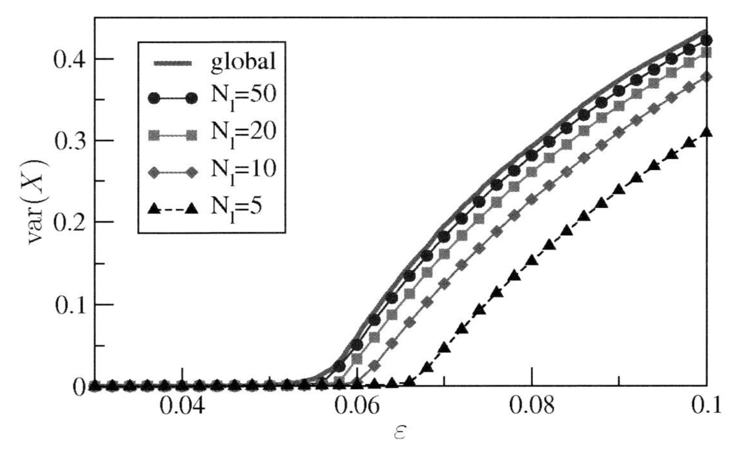

Fig. 6.2 Comparison of synchronization transition in ensembles with all-to-all coupling and random coupling, when each unit is coupled to N_l units. Bold line shows the transition for the case of all-to-all coupling. It is seen that the dynamics of the ensemble with global coupling is very close to the case of an ensemble with $N_l = 50$. For smaller number of links, the dynamics is qualitatively similar, though the synchronization transition happens for larger coupling ε. Each unit is modeled by Rulkov neuron map, see Eqs. (6.1). This simulation supports application of mean field approximation to not fully connected neural ensembles.

units. There are many possibilities to model a randomly coupled network; here we consider that each of $N = 10000$ neurons in Eq. (6.1) has N_l links, i.e. it is coupled to N_l randomly chosen elements of the population. The coupling strength ε within each pair is taken to be the same. In Fig. 6.2 we plot the dependence of the variance of the mean field $\text{var}(X)$ on ε for different number of links N_l. The synchronization transition is similar to the case of the global coupling, though it takes place for higher values of ε. We see that the bifurcation curve for $N_l = 50$ practically coincides with that for the global coupling case. Thus, the dynamics of an ensemble with random coupling to N_l neighbors is qualitatively close to the dynamics of the globally coupled population.

6.2
Why to Control?

The control problem we discuss in this chapter – manipulation of the collective mode of a large population of oscillators – is motivated by a hypothesis linking an important problem of neuroscience, namely suppression of pathological brain rhythms, to the just described phenomenon of synchronization in a large network of interacting oscillatory units. This hypothesis says that the well-pronounced rhythm, observed, e.g., in records of electrical and magnetic activity of Parkinsonian patients can be regarded as the mean field in a population of neurons [29]. Thus, emergence of a pathological rhythm was assumed to be a consequence of pathological synchronization in a neural ensemble. Correspondingly, the problem of suppression of this undesired activity was formulated as the problem of *desynchronization* [29] and a number of techniques has been proposed in order to tackle this problem. The ultimate goal of these theoretical studies is to substitute the *Deep Brain Stimulation (DBS)* – an empirical technique, currently used in clinical practice for suppression of tremor in Parkinson's disease [30] and for treatment of other syndromes [31] – by a more efficient and mild technique. The suggested approaches can be classified into two groups: non-feedback (see review in [32]) and feedback techniques [33–37].

In this review we consider the desynchronization problem from the physical standpoint and discuss advantages of feedback-based methods, concentrating on the techniques which allow one to desynchronize the system and to maintain the asynchronous state by a vanishing control input to the system. In biological terms, this would mean suppression of the tremor by a minimal stimulation of the neural tissue. Practical implementation of such a technique would reduce interference into the live system and, hopefully, reduce the side effects.

6.3 Controlling Neural Synchrony

In this section we discuss specific problems related to a control of a large populations of neurons.

6.3.1 Electrical Stimulation of Neural Ensembles

The first issue is to discuss a possibility to influence the spontaneously active, i.e. spiking and/or bursting, neurons. A common tool of experimental neuroscience is electrical stimulation of a neural tissue; it is frequently used in *in vitro* and *in vivo* animal experiments. For a couple of decades an electrical stimulation of the human brain via implanted electrodes aimed at quenching of pathological rhythms has been used in pilot studies and in clinical practice. To the best of our knowledge, the first attempt was reported in 1978 by S. A. Chkhenkeli [38, 39]. In this study the brain activity of epileptic patients was monitored and if the total spectral power in a certain range exceeded a chosen threshold level, indicating the onset of an epileptic seizure, the stimulation was switched on. With this demand-controlled stimulation Chkhenkeli managed to stop epileptic seizures. More successful was clinical applications of DBS in treatment of Parkinson's disease [30]. This procedure involves implantation of micro-electrodes into the subcortical structures and long-term stimulation by a periodic pulse train, typically with frequencies greater than 100 Hz. It has been shown that DBS relieves tremor as well as other symptoms such as rigidity and dyskinesia. It decreases tremor amplitude in a spectacular way and therefore is widely used nowadays in medical practice; so one of the producers of DBS controllers, the Medtronic Inc, reports on over 20 thousands of patients using their devices.

However, there exist some practical and theoretical problems. First, the mechanism by which high frequency DBS suppresses tremor and reduces other symptoms in Parkinson's disease is unknown and remains a subject of debates. Next, the parameters of the stimulation must be determined by trial and error and re-adjusted with time. The efficiency of the DBS is known to decrease with time due to the adaptation of the brain. Moreover, this technique is not free from side-effects. Permanent or long-term stimulation by high-frequency pulses is energetically high-consuming, and therefore the battery in the controller should be exchanged rather frequently; as the controller is implanted under the skin on the chest, every exchange requires another surgical intervention. Hence, the energy consumption of the device should be minimized. Finally, permanent stimulation definitely represents a very strong intervention into the system. All this calls for development of more intelligent stimulation techniques, which would be able to achieve suppression of a pathological brain activity by a minimized stimulation.

6.3.2
What Does the Control Theory Say?

Suppression of an undesired rhythm can be considered as a stabilization of an unknown unstable steady state of a complex multi-dimensional system. Stabilization of steady states is a classical problem of the control theory. A common approach to treat this problem is to implement a feedback control. Typically, the feedback signal is proportional to the deviation of a coordinate of the systems from the desired state (proportional control), or to the derivative of the coordinate (proportional-derivative control), or to the integral of the coordinate over the past (proportional-integral control), or to a combination of these three values [40]. Another group of stabilization techniques uses linear or nonlinear *time-delayed* feedback, see, e.g., [41–44]. For low-dimensional systems, the theory of feedback stabilization is well-developed and finds many technical applications.

Let us now discuss the feedback control from the view point of its possible applications to DBS. Clearly, a feedback scheme should possess a very important property: the control signal (the stimulation, in terms of neurophysiology) should vanish as soon as the goal of the control is achieved and the undesired rhythm is suppressed. So, e.g., the control, proportional to $\mathbf{x} - \mathbf{x}_0$, where $\mathbf{x}$ is the state vector of the system under control and $\mathbf{x}_0$ is the point to be stabilized, obviously possesses this property (note, however, that implementation of such a control requires knowledge of the fixed point $\mathbf{x}_0$). In context of chaos control, such a scheme is called *noninvasive*. We prefer here to call such a scheme a *vanishing-stimulation control*, because even for a vanishing stimulation, micro-electrodes should remain implanted, so that from the viewpoint of neuroscience, the technique certainly remains invasive. In the following we concentrate only on vanishing stimulation techniques.

In particular, this property can be easily implemented by a delayed-feedback control, if the control signal $\mathcal{C}$ is proportional to the difference between the current and delayed value of a system coordinate, i.e., if

$$\mathcal{C} \sim (x(t-\tau) - x(t)) \ ,$$

where τ is the delay. Obviously, if such a control suppresses oscillations around some fixed point x_0, then the control signal vanishes. Note, that no *a priori* knowledge of x_0 is required. Note also, that a control proportional to the delayed term, $\mathcal{C} \sim x(t-\tau)$ is non-vanishing, if $x_0 \neq 0$. This property becomes not so obvious for more complicated, e.g., nonlinear feedback control schemes [43, 44]. So, nonlinear feedback $\mathcal{C} \sim x^3(t-\tau)$ of the van der Pol oscillator near the Hopf bifurcation point is non-vanishing, though $x_0 = 0$: contrary to linear feedback that shifts the bifurcation point and therefore completely quenches the oscillation, the nonlinear feedback just reduces the oscillation amplitude [43,44] (cf. [35]).

Another way to organize a vanishing stimulation control in case of an unknown fixed point x_0 is to estimate this fixed point dynamically. This can be implemented by means of a high-pass first order filter in the feedback loop (washout filter) [45,46].

6.3.3 Control of Brain Rhythms: Specific Problems and Assumptions

The idea to implement the feedback control for suppression of pathological rhythms, suggested in our previous publications [33,34], is based on the assumptions that (1) the collective activity of many neurons is reflected in the local field potential (LFP) $\Phi(t)$ which can be registered by an extracellular microelectrode and (2) the processed and transformed LFP signal (control signal) can be fed back into the system via the second electrode. (Technically it is possible to perform simultaneous recording and stimulation using only one electrode, see [47] and references therein). The first assumption is supported by the knowledge of physical mechanisms of LFP generation [48,49]; it is known that the potential registered by an electrode in the inter-cellular space is $\Phi \sim \sum_i (\mathcal{I}_i / r_i)$, where r_i is the distance between the current source, i.e. the membrane current of the ith neuron $\mathcal{I}_i$, and the measuring point. Hence, in the first approximation, neglecting the spatial structure of the neural population, we can represent the measured signal as

$$\Phi \sim \sum_i \mathcal{I}_i \,. \tag{6.2}$$

Let us now discuss the second assumption that is common for all techniques intended for DBS. The detailed electro-physiological mechanisms of stimulation are largely unknown. It is not clear whether indeed the whole population or at least its large part can be affected in a common way. We assume that it is the case. However, there is an uncertainty in how the stimulation can be incorporated into equations, and we characterize this uncertainty by an unknown phase shift β, inherent to stimulation; below we discuss this quantity in more detail. Presence of the phase shift β puts forward a certain requirement to the control scheme: it should be able to compensate this shift, as well as the latency in measurements.

Another important requirement is as follows: the controller should be able to extract the relevant signal from its mixture with the rhythms produced by neighboring neuronal populations and with the measuremental noise. Below we describe possible solutions of the formulated problem.

6.4
Delayed Feedback Control

An application of delayed feedback control has become quite popular since publication of Pyragas' paper on stabilization of periodic orbits in chaotic systems [41]. This method has been studied quite intensively [50], moreover it has been recently extended to a problem of a control of chaos coherence [51]. As this method can be easily implemented experimentally (provided the characteristic time scales are not too small), it appears promising to apply it to a control of collective synchrony.

6.4.1
An Example

We start a discussion of delayed feedback suppression by considering an example, where we simulate the dynamics of an ensemble of 10000 Hindmarsh–Rose neurons coupled via a mean field. Individual neurons are in the regime of chaotic bursting. The equations of the system read:

$$\begin{aligned} \dot{x}_i &= y_i - x_i^3 + 3x_i^2 - z_i + 3 + \varepsilon X + \varepsilon_f (X(t-\tau) - X(t)) \,, \\ \dot{y}_i &= 1 - 5x_i^2 - y_i \,, \\ \dot{z}_i &= 0.006 \cdot (4(x_i + 1.56) - z_i) \,, \end{aligned} \tag{6.3}$$

where $X = N^{-1} \sum_{i=1}^{N} x_i$ is the mean field, and the terms εX and $\varepsilon_f(X(t-\tau) - X(t)) = C$ describe the global coupling and the feedback control, respectively. It means that here we take the simplest model of the local field potential measurement, assuming $\Phi = X$; this assumption will be reconsidered below. The results of simulation are shown in Fig. 6.3 for the following values of parameters: $\varepsilon = 0.08$, $\tau = 72.5$. The control is switched off for time $t < 5000$ and switched on with $\varepsilon_f = 0.036$ for $t > 5000$. Note that here we use the simplest model of the local field potential measurement, taking $\Phi(t) = X$; this issue will be re-discussed below.

Figure 6.3 illustrates two main properties of our suppression scheme. First, it implements suppression only on the macroscopic level, while individual neurons continue bursting, as before, just not coherently. Second, the control term decreases very rapidly and remains small, fluctuating around zero. Note that the mean field itself fluctuates around a constant level ≈ -1. However, due to the differential character of the control term, $C \sim (X(t-\tau) - X(t))$, the control input to the system vanishes.

The illustrated suppression scheme has three parameters. Two parameters are the feedback strength ε and the delay time τ. Numerical studies show that the suppression is effective in rather large domains in the plane (ε, τ).

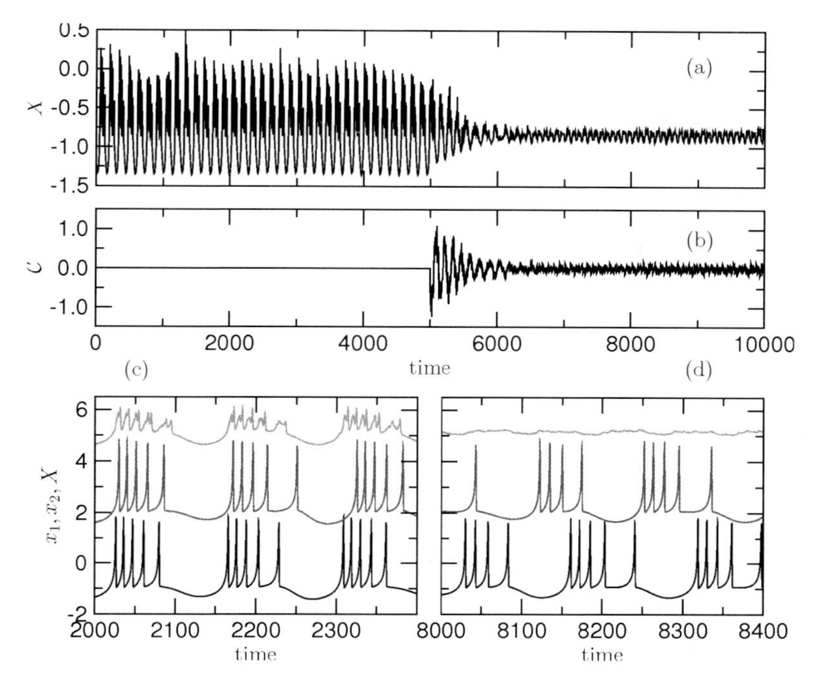

Fig. 6.3 (a) Mean field in an ensemble of 10000 Hindmarsh–Rose neurons without ($t < 5000$) and with ($t > 5000$) an external feedback. Without feedback, the mean field is large, indicating a certain level of synchrony between individual units. This is illustrated by panel (c), where the dynamics of two neurons and of the mean field are shown (the curves are shifted in the vertical direction for presentation purposes). Note that neurons are synchronized on the slow time scale, i.e. they burst in a coherent fashion (though this coherence is not perfect), and not synchronized on the fast time scale, i.e. in spiking. This results in irregularity of the mean field. (b) The control signal is of the same order of magnitude as the mean field for a rather short time after the feedback is switched on, and then quickly decays to the noise level. Panel (d) illustrates the main property of our control: the activity of individual neurons is not suppressed, but just desynchronized.

These domains are found around values of delay $\tau = \text{const} + nT/2$, where T is the period of mean field oscillation without control, and n is an integer. The constant here depends on the third parameter, that was not given explicitely in the Eqs. (6.3). This parameter, the phase shift β, is inherent to stimulation, and depends on how the stimulation enters the system equations; this parameter will be discussed below.

6.4.2 Theoretical Description

There are two approaches to the theoretical analysis of the delayed feedback control of collective synchrony. One follows [52,53] and uses noisy phase oscillators models and Fokker–Planck description for the probability distribution of phases ρ [33]. The analysis of stability of the uniform solution $\rho = 1/2\pi$ provides the domains where the control is effective.

The second approach is based on the assumption that the mean field appears via a Hopf bifurcation. Hence, the dynamics of the collective mode can be described by the following model equation:

$$\dot{A} = (\xi + i)A - \eta|A|^2 A + \varepsilon_f e^{i\beta}(A(t-\tau) - A)\,, \tag{6.4}$$

where the last term describes differential form of the feedback control. Linearization and analysis of stability of the fixed point solution $A = 0$ (what corresponds to an asynchronous state of the population) yields the characteristic equation

$$\lambda = \xi + i + \varepsilon_f e^{i\beta}(e^{-\lambda\tau} - 1)\,, \tag{6.5}$$

which can be solved (see [34]) in terms of the Lambert function [54]. The lines $\text{Re}(\lambda(\tau, \varepsilon_f)) = 0$ provide the borders of the stability domains in the plane (τ, ε_f), i.e. domains where the control is efficient (Fig. 6.4). This simplistic ap-

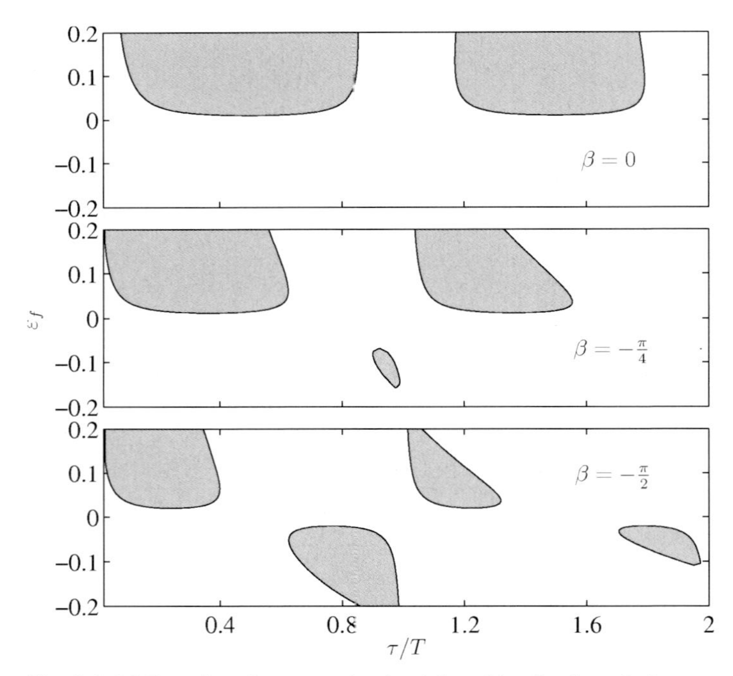

Fig. 6.4 (a) Domains of suppression for delayed feedback control, for different values of the stimulation phase shift β. The domains are obtained by solving Eq. (6.5); in this example $\xi = 0.02$. The delay τ is normalized by the mean period T of the mean field without control.

proach provides, however, a good correspondence with the numerical analysis of ensemble models, see Ref. [34] for details. In spite of this, many aspects which are important from the viewpoint of application can be analyzed only numerically. These aspects are considered in the following section.

6.4.3 Beyond Idealized Model

A theoretical treatment is possible only for infinitely large ensembles with full connectivity. Furthermore, our analysis above implied that the mean field can be measured perfectly and that all elements of the population can be equally influenced by the stimulation. These assumptions are certainly not realistic, and here we briefly present results of numerical analysis of non-ideal models [36], addressing several practical issues. For this purpose we introduce a qualitative measure of suppression efficacy, which we call the suppression factor

$$S = \left(\frac{\text{var}(X)}{\text{var}(X_f)} \right)^{1/2}, \tag{6.6}$$

where $\text{var}(X)$ and $\text{var}(X_f)$ are the variance of the mean field in the absence and presence of the control, respectively.

1. **Full vs not full connectivity.** This issue was addressed in Fig. 6.2. This figure demonstrates that an ensemble with relatively low long-range connectivity, e.g. when each element is linked to 50 randomly chosen units out of 10 000, can be very well approximated by a mean field model. As neurons indeed have many connections to distant neighbors through long axons, the use of a mean field model is justified.

2. **Finite size effects.** Numerical analysis shows that the dependence of the suppression factor S on the population size N can be perfectly fit by the square root function, $S \sim \sqrt{N}$, in correspondence to theoretical considerations. Indeed, in the finite-size population the mean field below the synchronization threshold can be treated as a noise with the variance $\sim N^{-1}$ [55]. On the other hand, for $\varepsilon > \varepsilon_{cr}$, $\text{var}(X)$ does not practically depend on N and, at least for small sub-criticality, $\text{var}(X) \sim \varepsilon - \varepsilon_{cr}$ (provided the Hopf bifurcation for the mean field is the supercritical one). Hence, the maximal possible suppression $S \sim \sqrt{(\varepsilon - \varepsilon_{cr})N}$.

3. **Additive noise in measurement.** To model the practical situation when the mean field is measured with an error, we took the stored signal as $X_\tau = X(t - \tau) + \xi$, where ξ was taken as white Gaussian noise. Simulations show that the suppression technique is quite robust with respect to

measuremental noise. So, for the ensemble of bursting Rulkov neurons, the suppression factor for certain control parameters was $S = 24$ in the noise free case. For $\text{rms}(\zeta)/\text{rms}(X) \approx 0.1$ it reduced to $S \approx 22$; for very strong noise $\text{rms}(\zeta)/\text{rms}(X) \approx 0.5$ the suppression was still efficient, $S \approx 8$.

4. **Imperfect measurement.** Now we suppose that the recording electrode measures not the mean field of the whole ensemble, but the mean field of a subpopulation containing qN neurons, $q \leq 1$, i.e. the control term has the form $\varepsilon_f X_q(t - \tau)$, where $X_q = (qN)^{-1} \sum_1^{qN} x(j)$, and $j = 1, \ldots, qN$ are the indices of a (randomly) picked subpopulation. Using X_q as the feedback signal one can still control the synchrony in the ensemble. The impact of the parameter q turns out to be similar to the finite-size effect, i.e. $S \sim \sqrt{q}$.
5. **Imperfect action.** Now we consider that the radiated signal acts only on a subpopulation of the size qN, $q \leq 1$. Simulations show that this imperfect action is equivalent to a decrease of feedback factor ε_f, and, hence, can be compensated, unless q becomes very small.
6. **Case of two interacting neuronal ensembles.** Now we briefly discuss how the technique works in case when the ensemble can be considered as consisting of two non-overlapping populations: the first one is affected by the stimulating electrode, and the mean field of the second one is measured by the recording electrode. Moreover, we consider two cases, when the second population is by itself stable or passive. The latter case may be considered as a model of suppression with a measurement from a surface electrode. Analytical analysis as well as simulations [36] demonstrate that, though the domains of suppression shrink, the maximal efficiency of the control remain preserved.

6.5 Suppression with a Non-delayed Feedback Loop

The physical mechanism of the delayed-feedback control is rather simple: time delay provides a proper phase shift so that the stimulation acts on each neuron in anti-phase to the mean field, and in this way compensates the latter. However, a phase shift can be provided also without delay. Below we present a scheme that implements this idea using a passive linear oscillator in the feedback loop.

From the viewpoint of the control theory it is an extension of filter-based techniques [45,46], where a filter in the feedback loop estimates the fixed point dynamically and eventually stabilizes it. From the viewpoint of nonlinear dynamics, our method goes back to the classical analysis of an interaction of an active oscillator with a passive load (resonator). It is known, that for certain parameters the load can quench the active system. We use here this idea in order to desynchronize an active medium.

6.5.1 Construction of a Feedback Loop

We include in the feedback loop a linear passive oscillator driven by the measured signal (local field potential) $\Psi(t)$:

$$\ddot{u} + \alpha\dot{u} + \omega_0^2 u = \Psi(t) \ . \tag{6.7}$$

The frequency of the oscillator ω_0 is taken to be equal to the frequency ω of uncontrolled mean field oscillations; this frequency can be easily measured in an experiment. This means that the driven oscillator (6.7) is in resonance with the forcing (for a moment we can consider it as a harmonic one with the frequency ω) and the phase of the output signal u is shifted by $\pi/2$ with respect to the phase of the input $\Psi(t)$, whereas the phase shift of the derivative of the output signal $\dot{u}$ with respect to the input $\Psi(t)$ is zero. Suppose for a moment that no additional phase shift is introduced by stimulation. Then control signal $\varepsilon\dot{u}$ would be in anti-phase with respect to $\Psi(t)$, if we choose $\varepsilon < 0$, and, hence, would counteract $\Psi(t)$, reducing synchrony. It is important to note that the variable $\dot{u}$ does not contain a constant component, $\langle\dot{u}\rangle = 0$, even if the observed field does. Thus, a stimulation proportional to $\dot{u}$ vanishes as soon as the control is successful, and the main requirement to the control strategy is fulfilled.

The considered feedback loop possesses one very useful feature: the output $\dot{u}$ can be considered as an application of a band pass filter to the input signal Ψ, which filters out noise and other components outside of the vicinity of the main oscillation mode.

Now we recall that the stimulation is characterized by an unknown phase shift β that should be compensated to ensure the required phase relations between the mean field and stimulation. To perform this, we compliment the feedback loop by a phase-shifting unit, described by

$$\mu\dot{d} + d = \dot{u} \ . \tag{6.8}$$

For $\mu\omega \gg 1$ this unit operates as an integrator (with an additional multiplication by the factor $1/\mu$), whereas for $\mu \to 0$ its transfer function is 1. Hence, the output of system (6.8) has the same average as the input $\dot{u}$, i.e. $\langle d\rangle = 0$. Finally, the control signal C is taken proportional to the weighted sum of $\dot{u}$ and d: $C \sim \varepsilon_f(\dot{u} + \gamma d)$, where the free parameter γ determines the desired phase shift. The units performing this summation and the integration according to Eq. (6.8) form the *phase shifter*. The phase difference θ between the output

$\dot{u} + \gamma d$ of the phase shifter and its input $\dot{u}$ is

$$\theta = -\arctan\left(\frac{\gamma}{\omega\mu}\right) , \tag{6.9}$$

and therefore can be arbitrary varied in the interval $-\pi/2 < \theta < \pi/2$. The phase shift in the interval $\pi/2 < \theta < 3\pi/2$ can be obtained by the sign inversion: $\varepsilon_f \to -\varepsilon_f$. With account of the phase shifter the equation for the control signal (stimulation) reads

$$C = \pm\frac{\varepsilon_f}{\sqrt{1+\gamma^2/\omega^2\mu^2}}(\dot{u} - \gamma d) = \varepsilon_f \cos\theta \cdot (\dot{u} - \omega\mu d \tan\theta) , \tag{6.10}$$

where $\sqrt{1+\gamma^2/\omega^2\mu^2} = 1/\cos\theta$ is the normalization coefficient. It ensures an independence of the amplification in the feedback loop from the phase shift θ, so that this amplification is completely determined by ε_f. Finally, we obtain the following system of equations for the collective mode:

$$\begin{aligned} \dot{A} &= (\xi + i\omega)A - |A|^2 A + \frac{\varepsilon_f}{\sqrt{1+\gamma^2/\mu^2\omega^2}}(\dot{u} + \gamma d)e^{i\beta} , \\ \ddot{u} + \alpha\dot{u} + \omega^2 u &= \mathrm{Re}(A) , \\ \mu\dot{d} + d &= \dot{u} . \end{aligned} \tag{6.11}$$

The desired, asynchronous state of the ensemble corresponds to the fixed point solution $A = 0$ of this system. The stability analysis, performed in [37],

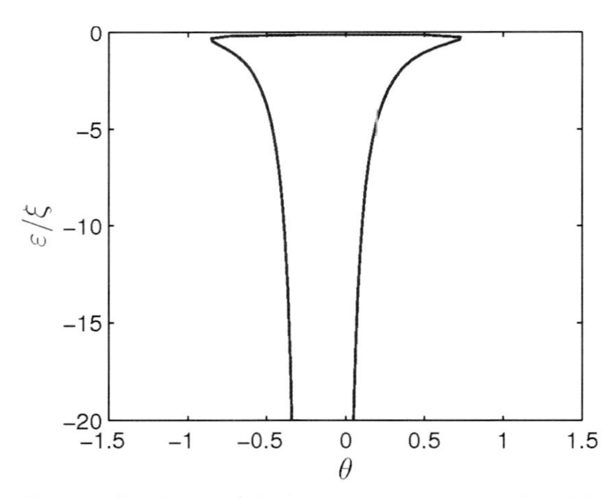

Fig. 6.5 Stability domain for the model equation (6.11) for $\beta = 0$ and $\xi = 0.0048$. The domain represents a large, though closed region, which extends to large negative values of ε_f. Note that only the area of strong stability is shown here. For different values of β the domain is shifted along the θ-axis.

provides domains of efficient control, which correspond to those obtained in simulation of ensemble models of different complexity (see Fig. 6.5 and cf. Fig. 6.6).

6.5.2
Example: Desynchronization of a Neuronal Ensemble with Synaptic Coupling

In the following example we take into account several important features of the measurement of the collective neuronal activity and of the coupling between the neurons. First, let us recall that the measured local field potential is proportional to the sum of the membrane currents $\mathcal{I}_i$, see Eq. (6.2), where $\mathcal{I}_i$ is the right hand side of an equation for the membrane potential V_i of a conductance-based neuronal model

$$C_i \frac{dV_i}{dt} = \mathcal{I}_i \, ,$$

where C_i is the capacitance of the membrane. Note that $\mathcal{I}_i$ are the total membrane currents which contain the currents through different ion channels and external currents, including the current due to stimulation. This means, that $\Phi \sim \sum_i \mathcal{I}_i = \sum_i \dot{V}_i$. Next, we note that interaction via the electrical coupling (gap junction) is possible only if the neurons are spatial neighbors. (The electrical coupling is a usual resistive coupling; it is possible only if the interacting neurons are closely spaced.) Therefore, in a large network, where even spatially distant neurons can be synaptically linked by long axons, synaptic coupling plays a more important role.

In view of this we explore the efficacy of suppression of a collective rhythm in a neuronal ensemble with all-to-all synaptic connections, where each neuron is modeled by the Hindmarsh–Rose equations [56]. The model and parameters of the inhibitory synaptic coupling are taken from Ref. [57]. Thus, the dynamics of the ensemble is described by the following set of equations:

$$\begin{aligned} \dot{x}_i &= y_i + 3x_i^2 - x_i^3 - z_i + I_i \, , \\ &\quad - \frac{\varepsilon}{N-1}(x_i + V_c) \textstyle\sum_{j \neq i}^{N} \left[1 + \exp\left(\frac{x_j - x_0}{\eta} \right) \right]^{-1} + \mathcal{C} \, , \\ \dot{y}_i &= 1 - 5x_i^2 - y_i \, , \\ \dot{z}_i &= r[\nu(x_i - \chi) - z_i] \, , \end{aligned} \tag{6.12}$$

where $r = 0.006, \nu = 1, \chi = -1.56$. ε is the strength of the synaptic coupling with the reverse potential $V_c = 1.4$; other parameters of synapses are $\eta = 0.01$, $x_0 = 0.85$. I_i is taken as $I_i = 4.2 + \sigma$, where σ is Gaussian distributed with zero mean and 0.05 rms value. For zero coupling, each neuron exhibits

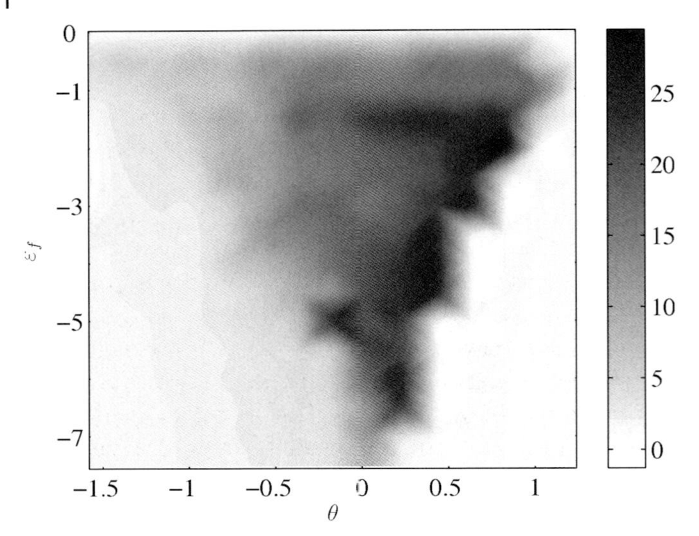

Fig. 6.6 Domain of suppression for the ensemble of 200 nonidentical synaptically coupled Hindmarsh–Rose neurons (Eq. (6.12)) in a regime of periodic spiking.

regular spiking. With the increase of the synaptic coupling between the neurons, the model demonstrates a transition from independent firing to coherent collective activity.[1)]

In the Eqs. (6.12) the variable x has the meaning of the membrane potential V. Hence, the local filed potential should be taken as $\Phi \sim \sum_i \dot{x}_i \sim \dot{X}$. It means, that the derivative of the mean field is measured and that the stimulation is now proportional to $\dot{X}$. While modeling the suppression by feedback control, we assume that the stimulation can be described as an additional external current, identical for all neurons. Therefore, the term, describing stimulation, enters the right hand side of the first Eq. (6.12).

The results of simulation for $N = 200$ nonidentical inhibitory coupled neurons are illustrated in Fig. 6.6. Here we show in gray-scale coding the suppression factor S as a function of the feedback strength ε_f and the phase shift θ. We remind that this shift is a free parameter of the control scheme and is intended to compensate the unknown phase shift β, inherent to stimulation. One can see that the suppression domain qualitatively agrees with the theoretical result (see Fig. 6.5). The simulations have been done for for the following values of the parameters: $\varepsilon = 0.15$ and $\alpha = 0.3\omega$. The average frequency of the mean field was estimated as $\omega = 2\pi/3.82$. Note that in this model the mean action

1) The dynamics of the coherent activity is quite complicated; it is likely that the system is in a self-organized quasiperiodic state (see [22]). However, it can be suppressed.

on each element *is not the mean field* X. Nevertheless, the measurement of $\dot{X}$ suffice to ensure desynchronization in the ensemble.

Finally, we consider the case when individual neurons exhibit chaotic bursting, i.e. generation of action potentials (spikes) alternates with the epochs of quiescence, so that the oscillation can be characterized by two time scales. This dynamics is provided by the Eq. (6.12a) with the following set of parameters: $r = 0.006, \nu = 4, \chi = -1.6$; the parameters of coupling are kept the same as in the previous example. Synchronization occurs on the slower time scale, i.e. different neurons burst nearly at the same time, whereas the spiking within the bursts is not synchronous, and therefore is to a large extent averaged out in the mean field (Fig. 6.7b). However, some high frequency jitter remains due to correlations in spiking. As a result, the mean field is irregular; besides this jitter, it also exhibits a low frequency modulation. The (average) frequency of the mean field is $\omega = 2\pi/176$, this corresponds to the inter-burst intervals.

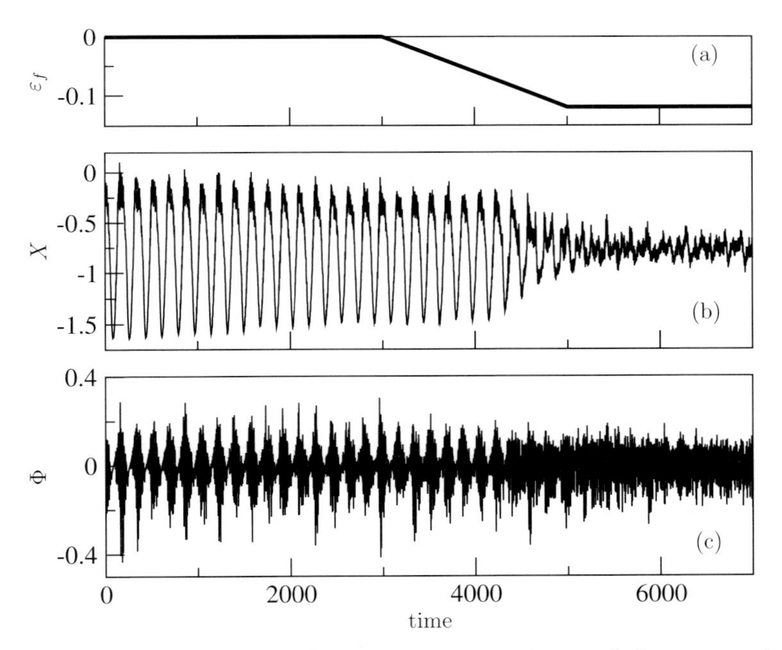

Fig. 6.7 (a) Feedback control ensures suppression of collective activity in an ensemble of bursting Hindmarsh–Rose neurons. Slow switching of the stimulation helps to avoid initial large current pulse. (b) Mean field in ensemble of synaptically coupled bursting Hindmarsh–Rose neurons is irregular but has a strong periodic component. The control has been switched on smoothly between $t = 3000$ and $t = 5000$. The suppression factor is $S = 6.5$, $\varepsilon_f = -0.12$, $\theta = -1.2$, $\alpha = 0.3\omega$. Note that for the measured signal we took the derivative of the mean field, shown in (c).

Figure 6.7(b) demonstrates that though the mean field is irregular, it has a strong periodic component and therefore we expect that our technique is efficient in this case as well. This is indeed confirmed by the results of numerical simulation for $I_i = 3.2$, $\varepsilon = 0.2$ and various values of the damping parameter α. We note that in order to avoid large current pulse at the beginning of the stimulation, the latter is switched on in a smooth way (Fig. 6.7a). As in the case of spiking neurons and in agreement with the theory, for the bursting neurons the stability domains elongate for rather large negative values of ε_f. Simulations also show that the increase of the damping parameter α, i.e. the increase of the bandwidth of the filter, leads to the extension of the suppression regions [37]. We conclude that suppression is possible in case of irregular mean field as well, as long as the latter has a strong periodic component (what is expected if the system is not too far from the point of transition to synchronization). We remind, that in order to simulate the measurement of LFP we use for stimulation the derivative of the mean field. This process (Fig. 6.7c) is even more complex than X; however, the suppression is achieved.

6.6
Determination of Stimulation Parameters by a Test Stimulation

As discussed above, determination of proper stimulation parameters requires a knowledge of the phase shift β. Here we show that β can be easily estimated by means of a test stimulation. Let us again consider the whole ensemble as one oscillator, writing the model amplitude equation for the collective mode

$$\dot{A} = (1 + i\omega) - |A|^2 A + e^{i\beta} C \,, \tag{6.13}$$

and try to identify β by applying an appropriate stimulation and observing the response. The main idea is to exploit the general synchronization property of oscillators. Suppose we stimulate the oscillator by a harmonic force which has the same frequency as the oscillator, i.e. $C = \varepsilon_f \cos(\omega t)$. If the phase shift β were zero, the oscillator should have the same phase as the force. Otherwise, the difference of the oscillator phase and phase of the force will be $\phi_{osc} - \phi_{force} = \beta$.

Hence, identification of the unknown parameter β can be performed in the following way: we first determine the frequency of the uncontrolled oscillation ω and then stimulate the ensemble by $\varepsilon_f \cos(\omega t)$. Computing the Fourier response on the frequency of stimulation, we get the estimate of β as

$$\tilde{\beta} = -\arg\left(T_s^{-1} \int_0^{T_s} \mathrm{Re}\, A e^{i\omega t} dt\right) \,, \tag{6.14}$$

where T_s is the time of stimulation.

To check this, we simulated the ensemble of 200 non-identical oscillators

$$\dot{A}_i = (1 + i\omega_i) - |A_i|^2 A_i + \varepsilon X + \varepsilon_f + e^{i\beta} \cos \omega_0 t \,, \tag{6.15}$$

where the frequencies ω_i are Gaussian distributed around 1 with the root mean square 0.05, and β was systematically varied from $-\pi$ to π. For each value of β we computed $\tilde{\beta}$ according to Eq. (6.15). Testing several values of ε_f, we found that the phase shift was reliably estimated with the help of weak test stimulation, $|\tilde{\beta} - \beta| < 0.2$.

6.7 Discussion and Outlook

In summary, we have reviewed techniques for the feedback suppression of collective synchrony in ensembles of oscillatory units. Having in mind possible application in neuroscience, we concentrated on methods providing the control with vanishing stimulation and thus minimizing the invasion into the system. We have shown, that this can be achieved by both delayed and non-delayed techniques. The latter can be easily implemented by incorporating a linear damped oscillator in the control loop. The theoretical analysis of suppression can be performed in the framework of the model amplitude equation; the results are in a good agreement with simulations.

An important advantage of the filter-based control is that this scheme has a built-in bandpass filter. This allows one to extract the relevant signal from its mixture with other rhythms and noise; the central frequency and the bandwidth of the filter are governed by parameters ω_0 and α (see Eq. (6.7)). With this method we also overcome the main disadvantage of the time-delayed method, namely that a new instability can arise if the delay is large enough.

Note that if the parameters are chosen in such a way that the stimulation acts in phase with the mean field, then both methods enhance the collective synchrony, what may be a useful feature for other applications, e.g. for creation of an experimental model of the disease. On the other hand, this might shed light on the emergence of pathological rhythms. Indeed, an intact brain has many natural feedback loops with delay. Then it might happen that a variation of parameters of such a loop can induce synchrony in an otherwise asynchronous population. In this case an appropriate approach to suppression would be to act on this loop, not on the population itself.

So far, the ideas presented in this review have been tested with rather simple, conceptual models of neural rhythm generation. The most complicated model, used in this context [36] consisted of two subpopulation of inhibitory and excitatory neurons with synaptic coupling, where individual cells were modeled by a discrete map [58]. However, this model describes only the dynamics of an isolated population of neurons. Modeling of the whole neural circuitry involved in the generation of the Parkinsonian rhythm remains a challenge. To our knowledge, the only step in this direction has been done by Rubin and Terman [59]. It is important, that their model can explain the effect of standard high-frequency deep brain stimulation, contrary to all other models used to test more advanced suppression techniques. However, the model of Rubin and Terman is rather complicated and computationally demanding, what hampers its application.

The ability of the delayed feedback control to manipulate the ensemble synchrony was confirmed in a recent experiment with electro-chemical oscillators [60]. However, verification of the applicability of these ideas to deep brain stimulation requires controlled experiments with neural cultures or slices, where both measurements of collective activity and of individual neurons are possible.

References

1 A. Pikovsky, M. Rosenblum, and J. Kurths. *Synchronization. A Universal Concept in Nonlinear Sciences*. Cambridge University Press, Cambridge, 2001.

2 Ch. Huygens (Hugenii). *Horologium Oscillatorium*. Apud F. Muguet, Parisiis, France, 1673. English translation: *The Pendulum Clock*, Iowa State University Press, Ames, 1986.

3 M. Bennett, M. Schatz, H. Rockwood, and K. Wiesenfeld. Huygens' clocks. *Proc. Royal Soc. (A)*, 458(2019):563–579, 2002.

4 J. Pantaleone. Synchronization of metronomes. *Am. J. Phys*, 70:992, 2002.

5 M. Bennett and K. Wiesenfeld. Averaged equations for distributed Josephson junction arrays. *Physica D*, 192(3-4):196–214, 2004.

6 H. Bruesselbach, D. C. Jones, M. S. Metin, and J. L. Rogers. Self-organized coherence in fiber laser arrays. *Optics Letters*, 30(11):1339–1341, 2005.

7 L. Glass. Synchronization and rhythmic processes in physiology. *Nature*, 410:277–284, 2001.

8 Z. Néda, E. Ravasz, Y. Brechet, T. Vicsek, and A.-L. Barabási. Tumultuous applause can transform itself into waves of synchronized clapping. *Nature*, 403(6772):849–850, 2000.

9 S. H. Strogatz, D. M. Abrams, A. McRobie, B. Eckhardt, and E. Ott. Theoretical mechanics: Crowd synchrony on the Millennium Bridge. *Nature*, 438:43–44, 2005.

10 Y. Kuramoto. *Chemical Oscillations, Waves and Turbulence*. Springer, Berlin, 1984.

11 J. A. Acebron, L. L. Bonilla, C. J. Perez Vicente, F. Ritort, and R. Spigler. The Kuramoto model: A simple paradigm for synchronization phenomena. *Rev. Mod. Phys.*, 77(1):137–175, 2005.

12 V. Hakim and W. J. Rappel. Dynamics of the globally coupled complex Ginzburg-Landau equation. *Phys. Rev. A*, 46(12):R7347–R7350, 1992.

13 K. Okuda. Variety and generality of clustering in globally coupled oscillators. *Physica D*, 63:424–436, 1993.

14 N. Nakagawa and Y. Kuramoto. Collective chaos in a population of globally coupled oscillators. *Prog. Theor. Phys.*, 89(2):313–323, 1993.

15 N. Nakagawa and Y. Kuramoto. From collective oscillations to collective chaos in a globally coupled oscillator system. *Physica D*, 75:74–80, 1994.

16 S. K. Han, C. Kurrer, and Y. Kuramoto. Dephasing and bursting in coupled neural oscillators. *Phys. Rev. Lett.*, 75:3190–3193, 1995.

17 K. Y. Tsang, R. E. Mirollo, S. H. Strogatz, and K. Wiesenfeld. Dynamics of globally coupled oscillator array. *Physica D*, 48:102–112, 1991.

18 H. Daido and K. Nakanishi. Diffusion-induced inhomogeneity in globally coupled oscillators: Swing-by mechanism. *Phys. Rev. Lett.*, 96:054101, 2006.

19 C. van Vreeswijk. Partial synchronization in populations of pulse-coupled oscillators. *Phys. Rev. E*, 54(5):5522–5537, 1996.

20 P.K. Mohanty and A. Politi. A new approach to partial synchronization in globally coupled rotators. *J. Phys. A: Math. Gen.*, 39(26):L415–L421, 2006.

21 A. Vilfan and T. Duke. Synchronization of active mechanical oscillators by an inertial load. *Phys. Rev. Lett.*, 91(11):114101, 2003.

22 M. Rosenblum and A. Pikovsky. Self-organized quasiperiodicity in oscillator ensembles with global nonlinear coupling. *Phys. Rev. Lett.*, 98:054102, 2007.

23 A. Pikovsky, M. Rosenblum, and J. Kurths. Synchronization in a population of globally coupled chaotic oscillators. *Europhys. Lett.*, 34(3):165–170, 1996.

24 D. Topaj, W.-H. Kye, and A. Pikovsky. Transition to coherence in populations of coupled chaotic oscillators: A linear response approach. *Phys. Rev. Lett.*, 87(7):074101, 2001.

25 E. Ott, P. So, E. Barreto, and T. Antonsen. The onset of synchronization in systems of globally coupled chaotic and periodic oscillators. *Physica D*, 173:29–51, 2002.

26 D. Golomb, D. Hansel, and G. Mato. Mechanisms of synchrony of neural activity in large networks. In F. Moss and S. Gielen, editors, *Neuro-informatics and Neural Modeling*, volume 4 of *Handbook of Biological Physics*, pages 887–968. Elsevier, Amsterdam, 2001.

27 N. F. Rulkov. Regularization of synchronized chaotic bursts. *Phys. Rev. Lett.*, 86(1):183–186, 2001.

28 N.F. Rulkov. Modeling of spiking-bursting neural behavior using two-dimensional map. *Phys. Rev. E*, 65:041922, 2002.

29 P. A. Tass. *Phase Resetting in Medicine and Biology. Stochastic Modelling and Data Analysis.* Springer-Verlag, Berlin, 1999.

30 A.L. Benabid, P. Pollak, C. Gervason, D. Hoffmann, D.M. Gao, M. Hommel, J.E. Perret, and J. De Rougemont. Long-term suppression of tremor by chronic stimulation of the ventral intermediate thalamic nucleus. *Lancet*, 337:403–406, 1991.

31 T. Wichmann and M. R. DeLong. Deep brain stimulation for neurologic and neuropsychiatric disorders. *Neuron*, 52:197–204, 2006.

32 P. A. Tass, Ch. Hauptmann, and O. Popovych. Development of therapeutic brain stimulation techniques with methods from nonlinear dynamics and statistical physics. *Int. J. Bif. & Chaos*, 16(7):1889, 2006.

33 M. G. Rosenblum and A. S. Pikovsky. Controlling synchrony in ensemble of globally coupled oscillators. *Phys. Rev. Lett.*, 92:114102, 2004.

34 M. G. Rosenblum and A. S. Pikovsky. Delayed feedback control of collective synchrony: An approach to suppression of pathological brain rhythms. *Phys. Rev. E*, 70:041904, 2004.

35 O. Popovych, Ch. Hauptmann, and P. A. Tass. Effective desynchronization by nonlinear delayed feedback. *Phys. Rev. Lett.*, 94:164102, 2005.

36 M. G. Rosenblum, N. Tukhlina, A. S. Pikovsky, and L. Cimponeriu. Delayed feedback suppression of collective rhythmic activity in a neuronal ensemble. *Int. J. of Bifurcation and Chaos*, 16(7):1989–1999, 2006.

37 N. Tukhlina, M. Rosenblum, A. Pikovsky, and J. Kurths. Feedback suppression of neural synchrony by vanishing stimulation. *Phys. Rev. E*, 75:011019, 2007.

38 S. A. Chkhenkeli. *Bull. of Georgian Academy of Sciences*, 90:406–411, 1978.

39 S. A. Chkhenkeli. Direct deep brain stimulation: First steps towards the feedback control of seizures. In J. Milton and P. Jung, editors, *Epilepsy as a Dynamic Disease*, pages 249–261. Springer, Berlin, 2003.

40 John Bechhoefer. Feedback for physicists: A tutorial essay on control. *Rev. Mod. Phys.*, 77:783, 2005.

41 K. Pyragas. Continuous control of chaos by self-controlling feedback. *Phys. Lett. A*, 170:421–428, 1992.

42 D. V. Ramana Reddy, A. Sen, and G. L. Johnston. Dynamics of a limit cycle oscillator under time delayed linear and nonlinear feedbacks. *Physica D*, 144(3-4):335–357, 2000.

43 F. M. Atay. Delayed-feedback control of oscillations in non-linear planar systems. *Int. J. Control*, 75:297–304, 2002.

44 F. M. Atay. Oscillation control in delayed feedback systems. In *Dynamics, Bifurcation, and Control*, volume 273 of *Lecture Notes in Control and Information Sciences*, pages 103–116. Springer-Verlag, Berlin, 2002.

45 M. A. Hassouneh, H.-C. Lee, and E. H. Abed. Washout filters in feedback control: Benefits, limitation and extensions. In *Proceeding of the 2004 American Control Conference*, pages 3950–3955. AACC. Boston, MA, 2004.

46 K. Pyragas, V. Pyragas, I.Z. Kiss, and J. L. Hudson. Adaptive control of unknown unstable steady states of dynamical systems. *Phys. Rev. E.*, 92:026215, 2004.

47 I. Ozden, S. Venkataramani, M. A. Long, B. W. Connors, and A. V. Nurmikko. Strong coupling of nonlinear electronic and biological oscillators: Reaching the "amplitude death" regime. *Phys. Rev. Lett.*, 93:158102, 2004.

48 U. Mitzdorf. Current source-density method and application in cat cerebral cortex: investigation of evoked potentials and eeg phenomena. *Physiological Reviews*, 65(1):37–90, 1985.

49 Peter beim Graben. *Symbolische Dynamik ereigniskorrelierter Gehirnpotentiale in der Sprachverarbeitung*. PhD thesis, Universität Potsdam, 2000.

50 H. G. Schuster, Editor. *Handbook of Chaos Control*. Wiley-VCH, Weinheim FRG, 1999.

51 D. Goldobin, M. Rosenblum, and A. Pikovsky. Controlling oscillator coherence by delayed feedback. *Phys. Rev. E*, 67:061119–7, 2003.

52 E. Niebur, H. G. Schuster, and D. M. Kammen. Collective frequencies and metastability in networks of limit-cycle oscillators with time delay. *Phys. Rev. Lett.*, 67(20):2753–2756, 1991.

53 M. K. S. Yeung and S. H. Strogatz. Time delay in the Kuramoto model of coupled oscillators. *Phys. Rev. Lett.*, 82(3):648–651, 1999.

54 R. M. Corless, G. H. Gonnet, D. E. G. Hare, D. J. Jeffrey, and D. E. Knuth. On the Lambert w function. *Advances in Computational Mathematics*, 5:329–359, 1996.

55 A. Pikovsky and S. Ruffo. Finite-size effects in a population of interacting oscillators. *Phys. Rev. E*, 59(2):1633–1636, 1999.

56 J. L. Hindmarsh and R. M. Rose. A model for neuronal bursting using three coupled first order differential equations. *Proc. Roy. Soc. London Ser. B*, 221:87, 1984.

57 R. Huerta, M. I. Rabinovich, H. D. I. Abarbanel, and M. Bazhenov. Spike-train bifurcation scaling in two coupled chaotic neurons. *Phys. Rev. E*, 55(3):R2108–R21010, 1997.

58 N. Rulkov, I. Timofeev, and M. Bazhenov. Oscillations in large-scale cortical networks: Map-based model. *J. Comp. Neurosci.*, 17:203–222, 2004.

59 J. Rubin and D. Terman. High frequency stimulation of the subthalamic nucleus eliminates pathological thalamic rhythmicity in a computational model. *J. Comp. Neurosci.*, 16:211–235, 2004.

60 J. Hudson et al. unpublished.

Index

Review of Nonlinear Dynamics and Complexity. Edited by Heinz Georg Schuster

ISBN: 978-3-527-40729-3